ZEITGEIST

AN ANTHOLOGY OF POPULAR SCIENCE WRITING

EDITORS
SIMON GAGE
PAULINE MULLIN

SUPPORTED BY

SCIENCE REVIEWS LTD

British Library Cataloguing in Publication Data
Zeitgeist
A catalogue record of this book is available from the British Library

ISBN 1-900814-05-6

Production editor: Sara Nash

Typeset by Cambridge Photosetting Services
Printed by Cambrian Printers, Aberystwyth

Science Reviews Ltd is the publishing arm of the Research Communications Foundation, an educational charity registered in England. The charity has English and American trustees and, unconstrained by commercial criteria, aims to encourage the publication of significant, scholarly texts which can also be enjoyed by those not expert in the field

Published by Science Reviews Ltd,
41–43 Green Lane, Northwood, Middlesex HA6 3AE, UK
Tel: 01923 823586 Fax: 01923 825066
E-mail: scitech@crac3.ch.kcl.ac.uk
web:http://crac3.ch.kcl.uk/scitech.htm

No part of this book may be reproduced by any mechanical, photographic, or electronic process or in any form of a phonographic recording, nor may it be stored in a retrieval system, transmitted or otherwise copied for public or private use without the written permission of the publisher.

© Science Reviews Ltd 1997

Contents

FOREWORD – *Kenneth Graham*	v
POLLEN GRAINS AND MAGIC BULLETS – *Richard Dawkins*	1
FUTURE ARCHAEOLOGIST – *Brian McCabe*	24
THIS SPORTING STRIFE – *Joe Collier*	25
THE CRANNOGS OF SCOTLAND – *Nicholas Dixon*	29
LIFE AS IT COULD BE – *Peter Coveney and Roger Highfield*	39
THERAPIST – *Norman Kreitman*	78
GIRLS (ONLY?) IN SCIENCE – *Martha C. Phelps-Borrowman*	79
MUSEUM – *Mary McCann*	94
BASING POLICY ON EVIDENCE AND PRINCIPLE: HOW TO INFLUENCE DECISION-MAKERS – *David Faulkner*	95
CRIMINAL JUSTICE AND PENAL POLICY IN SCOTLAND – *Peter Young*	111
ALIENS FROM SPACE? – *Patrick Moore*	113
NOT INVENTED HERE – *Iain M. Banks*	121
SCIENCE AND THE RETREAT FROM REASON – *John Gillott*	139
THE CHEMISTRY LESSON – *Ranald Macdonald*	148
WHAT MATHEMATICS IS FOR – *Ian Stewart*	149
OUT OF THIS WORLD – *Photographs by Kenny Bean*	161
ALTERNATIVES AND THE ETHICS OF USING ANIMALS IN RESEARCH – *Myc Riggulsford*	171
IMPURE TRUTH – *Tessa Ransford*	180
GENES, CANCER AND PREVENTION – *Ian Kunkler*	181
THE ENVIRONMENT: A MARRIAGE OF DISCIPLINES – *Crispin Tickell*	191
ICE AGE – *Ken Morrice*	200
THE EBB AND FLOW OF GEOLOGY IN BRITAIN – *Norman E. Butcher*	201
SCIENCE FESTIVAL (3.18×10^8 m s^{-1}) – *Michael Forester*	221

Acknowledgements

Richard Dawkins
Extract is a complete chapter from *Climbing Mount Improbable* by Richard Dawkins, published by Viking 1996 and Penguin 1997.

Joe Collier
This Sporting Strife by Joe Collier courtesy of The Guardian © 20th March 1996.

Peter Coveney and Roger Highfield
Extract from *FRONTIERS OF COMPLEXITY: The Search for Order in a Chaotic World*. Paperback edition published by Faber and Faber Ltd, 23 September 1996, £9.99.

Iain M Banks
Extract from *Excession* by Iain M Banks, published by Little Brown, London, 1996.

Ian Stewart
Extract from *Nature's Numbers* by Ian Stewart, published by Weidenfeld & Nicolson, 1995.

Illustrations for Norman Butcher's essay
Sing a Song of Symmetry courtesy of Professor David Daiches, in *A Weekly Scotsman and Other Peoms*, Black Ace Books, 1995.
James Hutton courtesy of The Scottish National Portrait Gallery.
Clerk of Eldin drawing of the Jedburgh Unconformity courtesy of Sir John Clerk of Penicuik.
Charles Lyell courtesy of Lady Lyell.
Temple of Serapis courtesy of John Murray (Publishers) Ltd, 50 Albemarle Street, London.
Section courtesy of John Murray (Publishers) Ltd, 50 Albemarle Street, London.
De la Beche courtesy of the British Geological Survey.
Schweppes advert courtesy of Cadbury Schweppes Ltd, Birmingham.
Subdivisions of Geology courtesy of Thomas Nelson and Sons Ltd.
Ian Glass courtesy of The Open University.
Understanding The Earth cover (S100) 1977, published by The Open University. Courtesy of The Open University.

Poetry
Impure Truth by Tessa Ransford from *Light of the Mind*, published by The Ramsay Head Press, 15 Gloucester Place, Edinburgh EH3 6EE.
Future Archaeologist by Brian McCabe from *One Atom to Another*, Polygon 1987.
Museum by Mary McCann in *Pomegranate*, Stramullion Publishing.
The Chemistry Lesson by Ranald Macdonald from *Lines Review* No 108 (p48), March 1989. New poems *The Naming of the Beasts* available from Chapman Publishing, 4 Broughton Place, Edinburgh EH1 3RX.
Therapist by Norman Kreitman from *Touching Rock*.
Ice Age by Ken Morrice from *For All I Know*, Aberdeen University Press 1981.

Sponsor
Zeitgeist is supported by The Post Office.

Others
With special thanks to:
Steven Mitchell, Tayburn McIlroy Coates
Tessa Ransford, Director, Scottish Poetry Library
Sir John Crofton

Foreword

THE POST OFFICE

Science, the attempt to understand why things are the way they are in an the immense diversity of this world and beyond, has always fascinated the curious and thoughtful. As we move towards the end of this century of unparalleled progress in science and technology, we find ourselves very much in the midst of an exciting period of step-change: that of instant information, itself a subset of the micro-electronics revolution that has gathered speed since the first transistor was developed nearly 50 years ago.

The information and micro-electronics revolutions will continue to impact on our leisure and work and therefore bring about change in society more pervasively, arguably, than other current scientific or technological activity.

The digital integration of telephone, television, satellite and computer offers an amazing communications potential, and the cyber-space world of the Internet promises access to almost unlimited quantities of information and the ability to publish and communicate on a global basis more easily than ever before.

In all this, the Postcode is the new geography; and this is a communications future the The Post Office is very much part of. Whether it is using Optical Character Recognition to help sort electronically the Postcodes of the massive 70 million letters handled every working day in the UK; using bar codes to keep track on packages anywhere in the world; or equipping the nation's largest retail chain – post offices – with on-line computers, The Post Office Group will continue to be an efficient provider of goods, services and information to business and the wider public in even the remotest communities. As always, science and technology must be seen as opportunities: not threats.

We think it relevant, therefore, to support this anthology published to co-incide with the 9th Edinburgh International Science Festival. If there is a unifying thread in both book and Festival it is communication. For it is communication which underpins successful science, just as it does success in the arts, and which forms the bridge to the wider world.

Even if all knowledge and progress are relative, scientific and technological change are irreversible. The excitement and the consequences are for everyone: not just the expert. If this book and the Festival help demonstrate that we shall be well pleased.

Kenneth Graham
Chairman, Scottish Post Office Board

Edinburgh International Science Festival

The Edinburgh International Science Festival aims to reveal science and the excitement of discovery to the widest possible audience. It provides a platform where scientists and the public can meet to discuss, debate and explore issues in science and technology which affect our everyday lives.

Since it began almost ten years ago, the Edinburgh International Science Festival has grown to become the largest public-access Science Festival in the world. Its innovative formula for presenting science and technology as part of a *festival* has already been copied by several countries around the world, many with the help and co-operation of the Edinburgh Science Festival.

The city of Edinburgh is a natural home for the Science Festival. It is internationally recognised as a seat of learning and a centre of excellence, particularly in the fields of science, technology, medicine and education, and is world renowned as *the* festival city.

The 1998 Festival, the Science Festival's tenth anniversary, runs from 4th to 19th April.

Edinburgh Science Festival, 149 Rose Street, Edinburgh, EH2 4LS, UK. Tel: 0131 220 3977 Fax: 0131 220 3987 e-mail: esf@scifest.demon.co.uk

Photo © Lisa Lloyd

Richard Dawkins is the first holder of Oxford's Charles Simonyi Chair of Public Understanding of Science, and a Professorial Fellow of New College, Oxford. His books, all best sellers and widely translated, are The Selfish Gene *(1976, 2nd Edition 1989)*; The Extended Phenotype *(1982, 1989)*; The Blind Watchmaker *(1986)*; River Out of Eden *(1995) and* Climbing Mount Improbable *(1996). This article is a complete chapter taken from* Climbing Mount Improbable *that was published by Viking in 1996 and Penguin in 1997. Richard Dawkins appears frequently on British television and radio, and in 1991 gave the Royal Institution Christmas Lectures for Children (televised by BBC). His prizes and awards include the Royal Society of Literature Prize, and the* Los Angeles Times *Literary Prize, the Silver Medal of the Zoological Society of London, the Michael Faraday Award of the Royal Society of London, and the Nakayama Prize for Human Science. He is married to the actress and artist Lalla Ward.*

POLLEN GRAINS AND MAGIC BULLETS

RICHARD DAWKINS

I was driving through the English countryside with my daughter Juliet, then aged six, and she pointed out some flowers by the wayside. I asked her what she thought wildflowers were for. She gave a rather thoughtful answer. 'Two things,' she said. 'To make the world pretty, and to help the bees make honey for us.' I was touched by this and sorry I had to tell her that it wasn't true.

My little girl's answer was not too different from the one that most adults, throughout history, would have given. It has long been widely believed that brute creation is here for our benefit. The first chapter of Genesis is explicit. Man has 'dominion' over all living things, and the animals and plants are there for our delight and our use. As the historian Sir Keith Thomas documents in his *Man and the Natural World*, this attitude pervaded medieval Christendom and it persists to this day. In the nineteenth century, the Reverend William Kirby thought that the louse was an indispensable incentive to cleanliness. Savage beasts, according to the Elizabethan bishop James Pilkington, fostered human courage and provided useful training for war. Horseflies, for an eighteenth-century writer, were created so 'that men should exercise their wits and industry to guard themselves against them'. Lobsters were furnished with hard shells so that, before eating them, we could benefit from the improving exercise of cracking their claws. Another pious medieval writer thought that weeds were there to benefit us: it is good for our spirit to have to work hard pulling them up.

Animals have been thought privileged to share in our punishment for Adam's sin. Keith Thomas quotes a seventeenth-century bishop on the point: 'Whatsoever change for the worse is come upon them is not their punishment, but a part of ours.' This must, one feels, be a great consolation to them. Henry More, in 1653, believed that cattle and sheep had only been given life in the first place so as to keep their meat fresh 'till we shall have need to eat them'. The logical conclusion to this seventeenth-century train of thought is that animals are actually eager to be eaten.

> The pheasant, partridge and the lark
> Flew to thy house, as to the Ark.
> The willing ox of himself came
> Home to the slaughter, with the lamb;
> And every beast did thither bring
> Himself to be an offering.

Douglas Adams developed this conceit to a futuristically bizarre conclusion in *The Restaurant at the End of the Universe*, part of the brilliant *Hitchhiker's Guide to the Galaxy* saga. As the hero and his friends sit down in the restaurant, a large quadruped obsequiously approaches their table and in pleasant, cultivated tones offers itself as the dish of the day. It explains that its kind has been bred to want to be eaten and with the ability to say so clearly and unambiguously: 'Something off the shoulder, perhaps? ... Braised in a white wine sauce? ... Or the rump is very good ... I've been exercising it and eating plenty of grain, so there's lots of good meat there.' Arthur Dent, the least galactically sophisticated of the diners, is horrified but the rest of the party order large steaks all round and the gentle creature gratefully trots off to the kitchen to shoot itself (humanely, it adds, with a reassuring wink at Arthur).

Douglas Adams's story is avowed comedy but, to the best of my belief, the following discussion of the banana, quoted verbatim from a modern tract kindly sent by one of my many creationist correspondents, is intended seriously.

> Note that the banana:
> 1. Is shaped for human hand
> 2. Has non-slip surface
> 3. Has outward indicators of inward contents: Green – too early; Yellow –just right; Black – too late
> 4. Has a tab for removal of wrapper
> 5. Is perforated on wrapper
> 6. Biodegradable wrapper
> 7. Is shaped for mouth
> 8. Has a point at top for ease of entry
> 9. Is pleasing to taste buds
> 10. Is curved towards the face to make eating process easy.

The attitude that living things are placed here for our benefit still dominates our culture, even where its underpinnings have disappeared. We now need, for purposes of scientific understanding, to find a less human-centred view of the natural world. If wild animals and plants can be said to be put into the world for any purpose – and there is a respectable figure of speech by which they can – it surely is not for the benefit of humans. We must learn to see things through non-human eyes. In the case of the flowers with which we began our discussion, it is at least marginally more sensible to see them through the eyes of bees and other creatures that pollinate them.

The whole life of bees revolves around the colourful, scented, nectar-dripping world of flowers. I am not just talking about honeybees, for there are thousands of different species of bee and they all depend utterly on flowers. Their larvae are fed on pollen, while the exclusive fuel for their adult flight-motors is nectar which is also entirely provided for them by flowers. When I say 'provided for them' I mean it in slightly more than an idle sense. Pollen, unlike nectar, is not provided *purely* for them, because the plants make pollen mainly for their own purposes. The bees are welcome to eat some of the pollen because they provide such a valuable service in carrying pollen from one flower to another. But nectar is a more extreme case. It doesn't have any other *raison d'être* than to feed bees. Nectar is manufactured, in large quantities, purely for bribing bees and other pollinators. The bees work hard for their nectar reward. To make one pound of clover honey, bees have to visit about ten million blossoms.

'Flowers', the bees might say, 'are there to provide us bees with pollen and nectar.' Even the bees haven't got it quite right. But they are a lot more right than we humans are if we think that flowers are there for our benefit. We might even say that flowers, at least the bright and showy ones, are bright and showy because they have been 'cultivated' by bees, butterflies, humming-birds and other pollinators. The original lecture upon which this chapter is based was called 'The Ultraviolet Garden'. This was a parable. Ultraviolet light is a kind of light that we can't see. Bees can, and they see it as a distinct colour, sometimes called bee purple. Flowers are bound to look very different through the eyes of bees (Figure 1). And in just the same way, the question 'What are flowers good for?' is a question that we are better off examining through the eyes of bees rather than through human eyes.

'The Ultraviolet Garden' plays on the strangeness of bee vision only as a parable for changing our point of view about who or what it is that flowers and all other living creatures – are 'for the good of'. If flowers had eyes, their view of the world might seem even odder to us than the alien ultraviolet visions of bees. How would bees appear through vegetable eyes? What are *bees* good for, from the point of view of the flowers? They are guided missiles for firing pollen from one flower to another. The background to this needs an explanation.

First, there are in general good genetic reasons for preferring cross-fertilization by pollen from a different plant. Incestuous self-mating would lose the benefits of sexual reproduction (whatever they are, which is an interesting question in itself). A tree that pollinated its female flowers with pollen from its own male flowers might almost as well not bother to pollinate at all. It would be more efficient to produce a vegetative clone of itself. Many plants of course do just this, and there is something to be said for it. But as we saw earlier there are also conditions in which there is even more to be said for reshuffling one's genes with those of another individual. It would require a massive digression to explain the detailed arguments, but there must be some substantial benefits to playing sexual roulette, otherwise natural selection wouldn't permit it to be such a driving obsession amongst almost all of animal and plant life. Whatever those benefits are, they would largely vanish if, instead of shuffling your genes with those of another individual, you simply shuffled them with a second, identical set of your own genes.

Flowers have no role in the life of their plant other than to exchange genes with another plant that has a different hand of genes. Some, like grasses, do it by wind. The air is lavishly flooded with pollen, a tiny proportion of which is lucky enough to drift on to the female parts of a flower of the same species (another proportion of it drifts into the noses and eyes of hayfever sufferers). This method of pollination is haphazard and, from some points of view, wasteful. It is often more efficient to exploit the wings and muscles of insects (or other vectors such as bats or humming-birds). This technique aims the pollen much more directly at its target, and consequently far less pollen is needed. On the other hand there has to be some expenditure on luring the insects. Part of the budget goes on advertising – bright-coloured petals and powerful scents. Part goes in bribes of nectar.

Nectar is high-quality aviation-fuel for an insect and it is costly for a plant to manufacture. Some plants duck out of the expense and employ deceptive advertising instead. Most famous are those orchids whose flowers look and smell like female insects. Male insects attempt to (Figure 1) copulate with the flowers and are inadvertently loaded

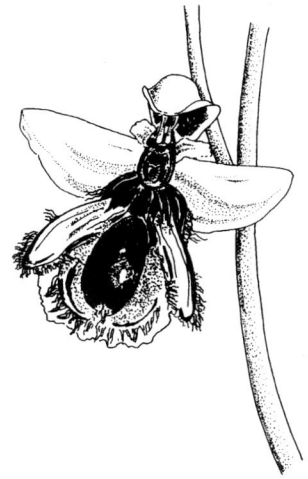

Figure 1 *Insect-mimicking orchid. Iberian Ophrys,* Ophrys vernixia.

with pollen bundles, or, at the other end of the trail, relieved of their pollen bundles. There are bee orchids that mimic female bees, and equally specialized fly orchids and wasp orchids. One of the wasp mimics, the well-named hammer orchid, keeps its dummy female wasp on the end of a hinged and spring-loaded stalk, cocked a fixed distance away from the pollen-bearing part of the flower (Figure 2). When the male wasp lands on the female dummy the spring is released. The male wasp is slammed, violently and repeatedly, against the anvil where the pollen sacs are kept. By the time the male wasp shakes himself free, his back is loaded with two pollen sacs.

Every bit as ingenious is the so-called bucket orchid, which works a little like a pitcher plant but with an important difference. The flower contains a large pool of liquid, alluringly scented to smell like the sexual attractant secreted by the females of a particular species of bee. A male of this species is attracted to the liquid, falls in and nearly drowns. The only escape is through a narrow tunnel. This the struggling bee eventually discovers and he crawls through it to salvation. At the far end of the tunnel there is a complicated gateway in which he is trapped for several minutes before he can wriggle free. During this final struggle at the portal of the tunnel, two large round pollen sacs are neatly transferred to his back. He then flies off and – sadder perhaps, but not wiser – falls into another bucket orchid. He again nearly drowns, again painfully pushes through the escape-tunnel and again is held for a while at the exit to freedom. During this period the second orchid relieves him of the pollen sacs and pollination is complete.

Never mind that 'sadder but not wiser'. As ever, the temptation to impute conscious intention should be resisted. It is, if anything, more tempting for the case of the plant. On both sides, the correct way to

Figure 2 *Hammer orchid*, Drakaea fitzgeraldii: *(a) the wasp alights on the lure; (b) the hinge buckles, slamming the wasp's back repeatedly against the pollinia.*

think of what is going on is in terms of unconsciously crafted machinery. Pollen that contains genes for building bee-manipulating bucket orchids is carried by bees. Pollen that contains genes for building orchids that are less accomplished at controlling bee behaviour is less likely to be carried by bees. So, as the generations go by, orchids get better at manipulating bees (although, actually, it has to be admitted that bee orchids are not in practice spectacularly successful at actually fooling bees into copulating with them).

These astonishing orchids epitomize an important aspect of pollination strategy. Many flowers seem to take great pains to get pollinated by one particular kind of animal but not any other. In the New World tropics, red tubular flowers are diagnostic of humming-bird pollination. Red is a bright and attractive colour to bird eyes (insects can't see red as a colour at all). Long, narrow tubes exclude all but specialist pollinators with long narrow beaks – humming-birds. Other flowers go out of their way to be pollinated only by bees, and we've already noted that their flowers are often coloured and patterned in the invisible (to humans) ultraviolet part of the spectrum. Yet others are pollinated only by night-flying moths. They are often white and they make use of scents in preference to visible advertisements. Perhaps the climactic stage in the progression towards an exclusive pollination partnership is the close hand-in-glove duo of fig trees with their own particular fig wasps, the example with which our book begins and ends. But why should plants be so fussy about who pollinates them?

Presumably the advantage of cultivating specialist pollinators is a more extreme version of the advantage of having animal pollinators at all, rather than wind. It narrows the target. Wind pollination is supremely extravagant, wastefully bathing the entire countryside in a rain of pollen. Pollination by jack-of-all-trades flying animals is better, but still pretty wasteful. The bee who visits your flower may fly on to a flower of a quite different species and your pollen will be wasted. Pollen borne by ordinary bees is not exactly rained over the countryside like that of a wind-pollinated grass, but it is still relatively indiscriminately splashed about. Contrast this with the bucket orchid's private species of bee, or a fig tree's private fig wasp. The insect flies unerringly, like a tiny guided missile, or like what medical journalists call a 'magic bullet', to exactly the right target from the point of view of the plant whose pollen it bears. In the case of the fig wasp, this means, as we shall see, not just another fig tree but another fig tree of precisely the right species out of the 900 fig species available. Employing specialist pollinators must permit huge savings in pollen production. On the other hand, as we shall also see, it raises other costs of its own, and it is not surprising that some plants are led by their way of life to stay with the wasteful wind as their pollinator. Other plant species are best suited by an intermediate technique along the spectrum from scattergun to magic bullet. Figs are perhaps the ultimate in dependency on the magic bullet of a particular species of pollinator and we reserve them for our climax, in the final chapter.

Returning to bees, the pollination services that they offer are truly massive. It has been calculated that, in Germany alone, honey-bees pollinate about ten trillion flowers in the course of a single summer day. It has also been calculated that 30 per cent of all human foods are derived from bee-pollinated plants, and that the economy of New Zealand would collapse if bees were wiped out. Bees, flowers might say, are put into the world to carry our pollen around for us.

The coloured and fragrant flowers of the world, then, although they may seem to be placed there for our benefit, are definitely not so. Flowers live in an insect garden, a mysterious ultraviolet garden in which, for all our vanities, we are irrelevant. Flowers have always been cultivated and domesticated but, until very recent times, the gardeners were bees and butterflies, not us. Flowers use bees, and bees use flowers. Both sides in the partnership have been shaped by the other. Both sides, in a way, have been domesticated, cultivated, by the other. The ultraviolet garden is a two-way garden. The bees cultivate the flowers for their purposes. And the flowers domesticate the bees for theirs.

Partnerships like this are quite common in evolution. There are so-called ant gardens consisting of epiphytes (plants that grow on the surface of other plants), which ants sow by bringing seeds of the right type and burying them in the soil of their nests. The plants grow out of the surface of the nest and their leaves provide food for the ants. It has been shown that some plants grow better if their roots are in an ants' nest. Other ant and termite species are specialized to cultivate fungi underground, planting the spores, weeding the gardens to rid them of competing fungus species, and fertilizing them with compost mulched from chewed-up leaves. In the case of the famous leafcutter ants of the New World tropics, all the foraging efforts of their eight-million-strong colonies are directed towards harvesting fresh-cut leaves. They can devastate an area with a ruthless efficiency reminiscent of a locust plague. Yet the leaves that they take are not to be eaten by the ants or their larvae, but are gathered purely to fertilize the fungus gardens. The ants themselves eat only the fungi, which are of a species that grows nowhere else than in the nests of this kind of ant. These fungi might say that ants are there purely to cultivate fungi, and the ants might say that the fungi exist purely to feed ants.

Perhaps the most remarkable of all the ant-loving plants are the South-East Asian epiphytes which grow a large bulbous swelling in the stem called a pseudo-bulb. The pseudo-bulb is hollowed with a labyrinth of cavities. These cavities are so like the ones ants commonly dig for themselves in soil that one would naturally suspect ants of fashioning them. This is not the case, however. The cavities are made by the plant and ants live in them (Figure 3).

Better known are the species of ants that live only in special hollow thorns of acacia trees (Inset Figure 4). The thorns are thick and bulbous and the plant makes them already hollowed out, apparently for no other purpose than to house ants. What the plants gain from the arrangement is protection, provided by the ants' vicious stings. This has been shown by elegantly simple experiments. Acacias whose ants have been killed by insecticide soon suffer marked increases in depredation from herbivores. Ants, if they think at all, think that acada thorns are for the benefit of ants. Acacias think that ants are for protecting them from browsers. Should we, then, think of each member of such partnerships as working for the good of the other? It is better to think of each as using the other for its *own* good. It is a kind of mutual exploitation in which each benefits from the other enough to make the costs of helping it worth paying.

There is a temptation, for which ecologists have been known to fall, to see all of life as a sort of mutual-support encounter-group. Plants are the community's primary energy harvesters. They trap the sun's rays and make its energy available to the whole community. They contribute to the community by being eaten. Herbivores, including

Figure 3 *A plant that provides custom-made accommodation for ants in return for protection. Cross-section of a pseudo-bulb of* Mynnecodia pentasperma.

the very abundant herbivorous insects, are the conduit by which the sun's energy is channelled from the primary producers, the plants, to higher stages in the food chain, the insectivores, small carnivores and large carnivores. When animals defecate or die, their vital chemicals are recycled by the scavengers such as dung beetles and burying beetles who hand the precious burden over to soil bacteria who eventually make it available to plants again.

There would not be too much wrong with this cosily benign picture of the circulation of energy and other resources, if only it were clearly understood that the participants are *not doing* it for the good of the circle. They are in the circle for the good of themselves. A dung beetle scavenges dung and buries it for food. The fact that she and her kind thereby perform a cleaning-up and recycling service which is valuable to the other inhabitants of the area is strictly incidental.

Grass provides the staple diet for a whole community of grazers, and the grazers manure the grass. It is even true that, if you removed the grazers, many of the grasses would die. But this does not mean that a grass plant exists to be eaten, or in any sense benefits by being eaten. A grass plant, if it could express its wishes, would much rather not be eaten. How, then, do we resolve the paradox that if the grazers were removed the grasses would die? The answer is that, although no plant wants to be eaten, grasses can tolerate it better than many other plants can (which is why they are used in lawns that are designed to be mown). As long as an area is heavily grazed or mown, plants that would compete with grasses cannot establish themselves. Trees cannot get a foothold because their seedlings are destroyed. Grazers, therefore, are indirectly good for grasses as a class. But this still does not mean that an individual grass plant benefits by being grazed. It may benefit from other grasses being grazed, including other plants of its own species, since this will have dividends in manure and in helping to remove competitor plants. But if the individual grass plant can get away with not being grazed itself, so much the better.

We began by lampooning the common fallacy that flowers and animals are placed in the world for the benefit of humans, cattle are docilely eager to be eaten, and so on. Marginally more defensible was the idea that they are placed in the world for the benefit of others with whom they have a naturally evolved mutualism: flowers for the benefit of bees, bees for the benefit of flowers, acacia bullhorns for the benefit of ants and their ants for the benefit of acacias. But this notion of creatures being 'for the good' of other creatures is in peril of *reductio ad absurdum*. We must have no truck with the pop ecologist's fallacy, the holisty grail of all individuals striving for the

Pollen grains and magic bullets

Figure 6 *Industrial robot from Nissan car factory, Yokohama.*

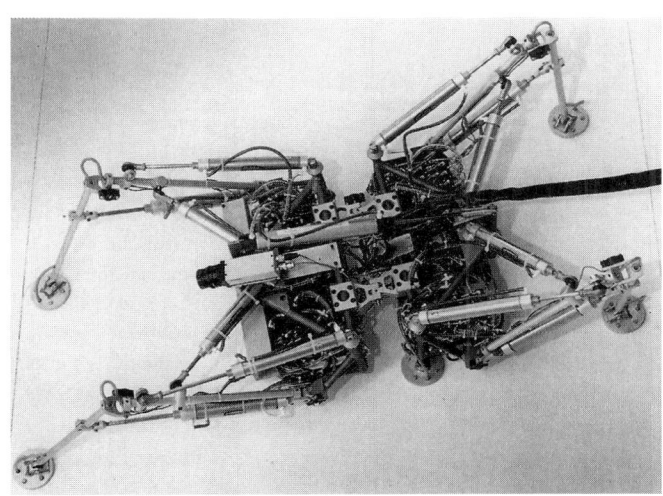

Figure 7 *Walking robot on sucker legs from Portsmouth Polytechnic, England*

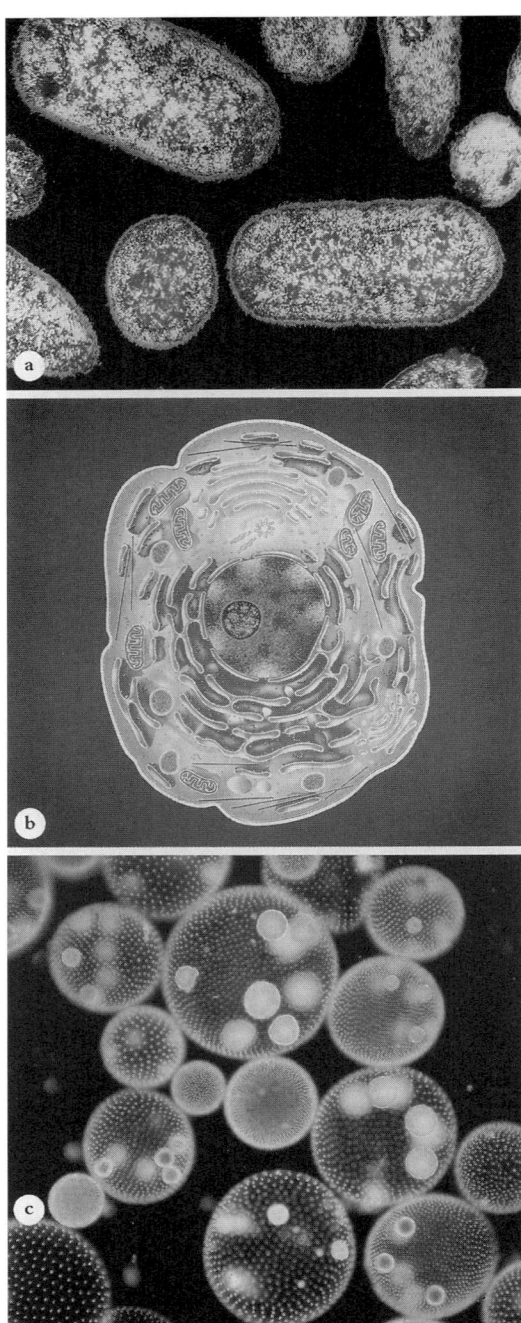

Figure 8 *Increasing levels of organization among life forms: (a) individual bacteria; (b) advanced – eucaryotic – cell with nucleus, originally evolved from a colony of bacteria; (c) volvox, a colony of eucaryotic cells; (d) a more densely packed and populous colony of differentiated eucaryotic cells, a tardigrade. A human body is another such colony – a colony of colonies, since each of our cells is a colony of bacteria; (e) a colony of indiviual organisms: a swarm of honey bees – a colony of colonies of colonies.*

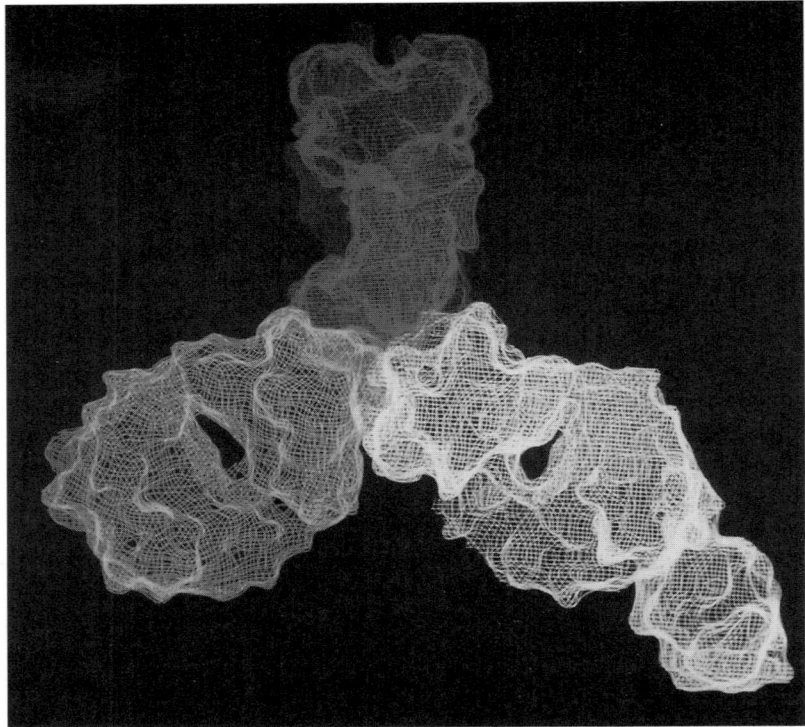

Figure 9 *Real life nanotechnology: an immunoglobulin molecule.*

good of the community, the ecosystem, 'Gaia'. It is time to get fussy and sharpen up what we mean whenever we talk of a living creature being there 'for the benefit of' anything. What does 'for the good of' really mean? What are flowers and bees, wasps and figs, elephants and bristlecone pines – what are all living things *really* for? What kind of an entity is it whose 'benefit' will be served by a living body or a part of a living body?

The answer is DNA. It is a profound and precise answer and the argument for it is watertight, but it needs some explanation. It is this explanation that I want to come on to now. I'll begin by returning to my daughter.

She was once suffering from a high fever and I suffered vicariously with her as I took my turns sitting by her bedside, sponging her down with cool water. Modern doctors could assure me that she was not in serious danger but the sleep-deprived mind of a loving father could not help recalling the countless childhood deaths of earlier centuries and the agony of each individual loss. Charles Darwin himself never

Pollen grains and magic bullets

Figure 10 *Interior of fig with male and female fig wasps.*

Figure 11 *The garden gate: exterior of fig to show the entrance.*

Pollen grains and magic bullets

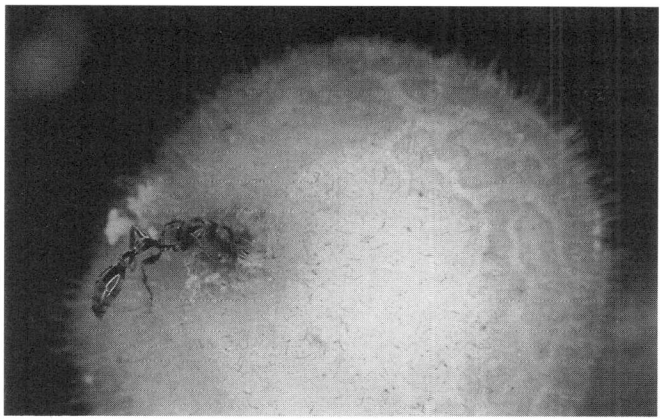

Figure 13 *Perils of being a fig wasp. An ant lurks outside the garden gate waiting for wasps to emerge.*

Figure 14 *Sectioned fig with parasitic wasp females waving their 'drilling rigs' in the air.*

Facing page: **Figure 12** *(a) strangler fig; (b) baobab tree entwined by strangler fig.*

17

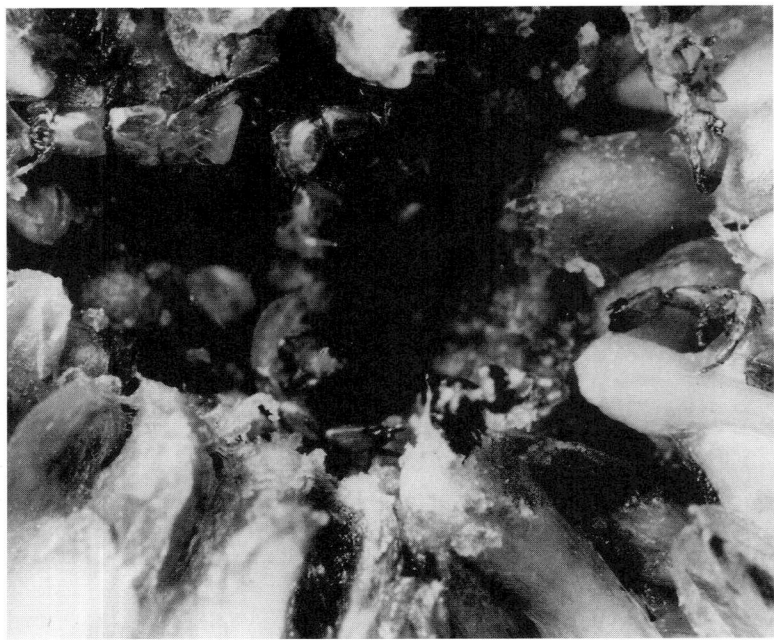

Figure 15 *A Garden Inclosed.*

recovered from the uncomprehended death of his beloved daughter Annie. The apparent injustice of her illness was said to have contributed to his loss of religious faith. If Juliet had turned to me and asked, in a piteous echo of our earlier and happier conversation, 'What are viruses for?', how should I have answered?

What are viruses for? To make us better and stronger through triumphing over adversity? (Like the 'benefits' of Auschwitz as was suggested by a professor of theology with whom I shared a debating platform on British television.) To kill enough of us to prevent the overpopulation of the world? (An especial boon in countries where effective contraception has been prohibited by theological authority.) To punish us for our sins? (In the case of the AIDS virus, you will find plenty of enthusiasts to agree. One feels almost sorry for medieval theologians that this admirably moralistic pathogen was not around in their time.) Once again, these replies are too human-centred, albeit in a negative way. Viruses, like everything else in nature, have no interest in humans, positive or negative. Viruses are coded program instructions written in DNA language, and they are for the good of the instructions themselves. The instructions say 'Copy Me and Spread Me

Around' and the ones that are obeyed are the ones that we encounter. That is all. That is the nearest you will come to an answer to the question 'What is the point of viruses?' It seems a pointless point, and that is precisely what I now wish to emphasize. I shall do so using the parallel case of computer viruses. The analogy between true viruses and computer viruses is extremely strong and it is also illuminating.

A computer virus is just a computer program, written in the same sort of language as any other computer program and travelling via the same range of media, for instance floppy discs, or the network of computers, telephone wires, modems and software that is called the Internet. Any computer program is just a set of instructions. Instructions to do what? It could be essentially anything. Some programs are sets of instructions to reckon accounts. Word processors are sets of instructions to accept typed words, move them around the screen and eventually print them. Yet other programs, like Genius 2 which recently defeated Kasparov, the Grand Master, are instructions to play chess very well. A computer virus is a program consisting of instructions that say something like this: 'Every time you come across a new computer disc, make a copy of me and put it on to that disc.' It is a 'Duplicate Me' program. It may incidentally say something more, for instance, 'Erase the entire hard disc.' Or it may cause the computer to speak, in tinny robotic tones, the words 'Don't panic'. But that is by the way. The hallmark of a computer virus, its identifying feature, is that it contains the instructions 'Duplicate me', written in a language that computers will obey.

Humans may see no reason to obey such starkly peremptory commands, but computers slavishly obey anything so long as it is written in their own particular language. 'Duplicate me' will be obeyed just as readily as 'Invert this matrix' or 'Italicize this paragraph' or 'Advance this pawn two squares'. Moreover, there is plenty of opportunity for cross-infection. Computer-users profligately exchange floppy discs, passing game programs around to friends, and useful programs too. You can easily see that, when there are lots of discs being promiscuously shared around, a program that said 'Copy me on to every disc you encounter, would spread around the world like chicken-pox. There would soon be hundreds of copies about, and the number would tend to increase. Nowadays, with information highways crisscrossing cyberspace, the opportunities for high-speed cross-infection by computer viruses are even better.

It is tempting to expostulate about the pointlessness of such parasitic programs, as I did when talking about disease viruses. What on earth is the use of a program that says nothing but 'Duplicate this program'? Admittedly it will be duplicated but isn't there something ridiculously otiose about such purely self-referential efforts? Of course there is! It is viciously futile. But it doesn't *matter* that it is futile and pointless in that sense. It can be utterly pointless and still spread. It spreads because it spreads because it spreads. The fact that it does nothing useful on the way – may even do something harmful on the way – is neither here nor there. In the world of computers and disc-swapping, it survives simply because it survives.

Biological viruses are just the same. Fundamentally a virus is just a program, written in DNA language, which is very much like a computer language even to the point of being written in a digital code. Like a computer virus, the biological virus simply says 'Copy me and spread me around'. As in the case of computer viruses we aren't suggesting that the DNA in a virus *wants to* get itself copied. It is just that, of all ways in which DNA could be arranged, only the arrangements that spell out the instructions 'Spread me' spread. The world willy-nilly becomes full of such programs. Once again, like the computer viruses, they're here because they're here because they're here. If they didn't embody instructions to ensure that they exist, they would not exist.

The only important difference between the two kinds of virus is that computer viruses are designed by the creative efforts of mischievous or evil humans, while biological viruses evolve by mutation and natural selection. If a biological virus has bad effects like sneezing or death, these are by-products or symptoms of its methods of spreading. The bad effects of computer viruses are sometimes of this type. The famous Internet Worm, which raced around the networks of the United States on 2 November 1988, had bad effects that were all non-deliberate by-products (a computer worm is technically distinct from a computer virus but the difference need not trouble us here). Copies of the program expropriated memory space and processor time, and brought around 6,000 computers to a standstill. Computer viruses, as we have seen, sometimes have bad effects which are not by-products or necessary symptoms but gratuitous manifestations of pure malice. Far from assisting the spread of the parasite these malicious effects, if anything, slow it down. Real viruses would do nothing so human-centred unless they were designed in a biological-warfare laboratory. Naturally evolved viruses don't go out of their way to kill us or make us suffer. They have no interest in whether we suffer or not. If we suffer, it is a by-product of their self-spreading activities.

'Duplicate me' instructions, like any instructions, are no use unless there is machinery set up to obey them. The world of computers is a fine and friendly place for a Duplicate Me program. Computers, linked by the Internet, abetted by people borrowing and lending discs, constitute a kind of paradise for a self-copying computer program. There is ready-made instruction-copying and instruction-obeying machinery humming and whirring and, in a sense, begging to be exploited by any program that says 'Duplicate me'. In the case of DNA viruses, the ready-made copying and obeying machinery is the machinery of cells, the whole elaborate paraphernalia of Messenger RNA, of Ribosomal RNA and of the various Transfer RNAs, each one hooking on to its own, key-coded amino acid. Never mind the details, or look them up in J. D. Watson's superbly clear *Molecular Biology of the Gene.* For our purposes it is enough to understand, first, that every cell contains a miniature analogue of a computer's instruction-obeying machinery and, second, that the machine code of all cells, in all creatures on Earth, is identical. (Computer viruses don't have that luxury, by the way: DOS viruses cannot infect Macs, and vice versa.) Computer virus instructions and DNA virus instructions are obeyed because they are written in a code that is slavishly obeyed in the environments in which they respectively find themselves.

But where does all this complaisant copying and instruction-executing machinery come from? It doesn't just happen. It has to be made. In the case of computer viruses, the machinery is made by humans. In the case of DNA viruses, the machinery is the cells of other creatures. And who manufactures those other creatures, those humans and elephants and hippos whose cells make life so easy for viruses? The answer is, other self-copying DNA manufactures them. The DNA that 'belongs' to the humans and the elephants. So, what *are* big creatures like elephants and cherry trees and mice? (I say 'big' because even a mouse, from a virus's point of view, is very very big.) And for whose benefit are mice and elephants and flowers put into the world?

We are closing in on a definitive answer to all questions of this kind. Flowers and elephants are 'for' the same thing as everything else in the living kingdoms, for spreading Duplicate Me programs written in DNA language. Flowers are for spreading copies of instructions for making more flowers. Elephants are for spreading copies of instructions for making more elephants. Birds are for spreading copies of instructions for making more birds. The cells of an elephant cannot tell whether the instructions they are slavishly obeying are virus instructions or elephant instructions. As in the case of Tennyson's Light Brigade when someone had blundered, 'Their's not to make reply, their's not to reason why, their's but to do and die'.

You will understand that I am using 'elephant' to stand for all large, autonomous creatures, for flowers or bees, for humans or cactuses, for bacteria even. The virus instructions, as we have seen, are saying 'Duplicate me'. What are the elephant instructions saying? This is the main insight that I wish to leave you with at the end of the chapter. Elephant instructions are also saying 'Duplicate me', but they are saying it in a much more roundabout way. The DNA of an elephant constitutes a gigantic program, analogous to a computer program. Like the virus DNA it is fundamentally a Duplicate Me program but it contains an almost fantastically large digression as an essential part of the efficient execution of its fundamental message. That digression is an elephant. The program says: 'Duplicate me by the roundabout route of building an elephant first.' The elephant feeds so as to grow; it grows so as to become adult; it becomes adult so as to mate and reproduce new elephants; it reproduces new elephants to propagate new copies of the original program instructions.

You can say the same about *bits* of creatures, too. The peacock's beak, by picking up food that keeps the peacock alive, is a tool for indirectly spreading instructions for making peacock beaks. The male peacock's fan is a tool for spreading instructions for making more peacocks' fans. It works by being attractive to peahens. It is good at picking up peahens while the beak is good at picking up food. Males with the most beautiful fans will have the most children to pass on copies of fan-beautifying genes. That is why peacock fans are so pretty. The fact that they are pretty to us is an incidental by-product. The peacock's fan is a gene spreader and it works via peahens' eyes.

Wings are tools for spreading genetic instructions for making wings. In the peacock's case they make their mark as gene preservers especially when the bird is surprised by a predator and shoots briefly into the air. Plants manage something akin to flight organs for their seeds (Figure 16), but in spite of this most people would probably not be happy to use the word 'flying', in its true sense, for plants. Plants, it seems, don't fly, and they don't have wings.

But wait! From a plant's point of view, it doesn't *need* wings of its own if it has bees' wings, or butterflies' wings, to do the job for it. In fact, I wouldn't mind *calling* the wings of a bee *plant wings*. They are organs of flight that are used, by the plant, to ferry its pollen from one flower to another. Flowers are tools for getting plant DNA into the next generation. They work like peacocks' fans, but instead of attracting peahens they attract bees. Otherwise there is no difference. Just as a peacock's fan works, indirectly, on the leg muscles of the peahen, causing her to walk towards the male and mate with him, so a flower's colours and stripes, its scent and its nectar, work on the wings of the bees and butterflies and humming-birds. The bees are drawn

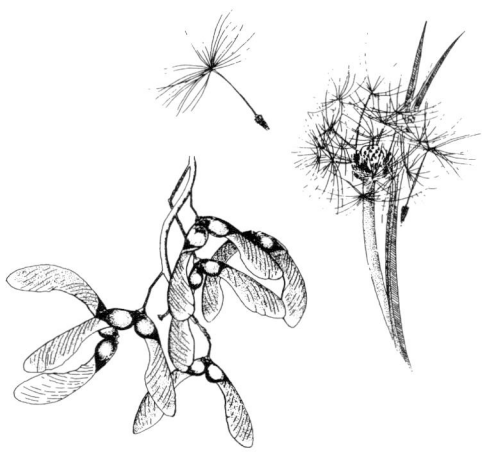

Figure 16 *DNA with wings: sycamore and dandelion seeds.*

towards the flowers. Their wings beat and carry the pollen from one flower to another. The wings of bees can truly be called flowers' wings, for they carry flower genes just as surely as they carry bee genes.

Elephant bodies cannot tell whether they are working to spread elephant DNA or virus DNA, and bees' wings cannot tell whether they are working to spread bee DNA or flower DNA. As it happens, if we set aside exceptional cases like the bees that are fooled into wasting their time copulating with bee orchids, they are working to spread both. The difference between 'own' DNA and pollen DNA, from the point of view of the bees' executive machinery, cannot be perceived. Peacocks and bees, flowers and elephants, stand to their own DNA in much the same relation as they stand to the DNA of parasitic viruses that infest them. Virus DNA is a program that says: 'Duplicate me in a simple and direct way, using the ready-made machinery of host cells.' Elephant DNA says: 'Duplicate me in a more complicated and roundabout way that involves, first, building an elephant.' Flower DNA says: 'Duplicate me in an even more complicated and roundabout way: first, build a flower and, second, use that flower to manipulate, by indirect influences such as seductive nectar, the wings of a bee (which has already conveniently been built to the specifications of another lot of DNA, the bee's "own" DNA) to carry far and wide the pollen grains inside which are the very same DNA instructions.'

FUTURE ARCHAEOLOGIST

What notion of yourself will you exhume
when you open that door in the stone
and step into our century's room?
Will our solemn, Tuttankhamun telephone
make you hear the echo of an instinct?
What will you do when you hear the tone
of our antiquated answering machine?
I'm out now. Please say who you are
and leave your message – I'll
call you back as soon as I can.
Amazed by what you have been,
will you tell us who you think you are?
Or will you say, 'But we've come so far
since then ...' Or, as our faces
vanish from the snapshot's catacomb,
'It's as if they are watching us ...'
don't mistake our carpeted rooms
for a shrine, our beds for our graves.
Future archaeologist, make it known
that we tried to love and be loved –
display our dark, embracing bones.

Brian McCabe

THIS SPORTING STRIFE

JOE COLLIER

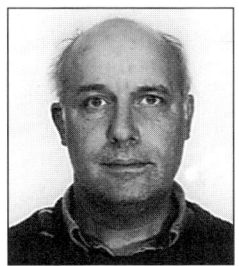

Joe Collier is Reader and Honorary Consultant in Clinical Pharmacology, editor of the Drug J. Therapeutics Bulletin *which is distributed to doctors throughout the UK, and is a specialist advisor on drugs to the House of Commons Health Select Committee. Apart from basic research he has a great interest in drug policy issues. He is the Clinical Secretary of his local Research Ethics Committee.*

One predictable outcome of the Commonwealth Games was the scandal about drug taking. The media – sports pages, commentators and leader writers – have overflowed with righteous outrage, the "cheats" have returned home under the cloud of orchestrated shame and the sports authorities have oozed self-satsified vindication for their policy. The organisers have, as on many previous occasions, defined a sports ethic, and woe betide those found in breach.

However, when it comes to laying down the law, the track record of sport's governing and organising bodies is shaky: why did they forbid full participation by women in so many fields; how could they permit women to be strip searched (sometimes by men) to check gender; what has happened to their stance on payment for taking part in competition? Sadly their current policy on drugs is in the same muddled tradition.

The problem starts with deciding who benefits from medicines, and whether or not this is acceptable. Sports and health organisations jointly promote more active lifestyles, and as competition becomes more widespread inevitably there will be those who have to take medicines to survive, let alone to compete. Drug taking certainly enhances the performance for the disabled or those who might otherwise be ill since, if deprived of medication, they could be no more than spectators. Rightly these competitors are not breaking rules, but where should the line be drawn? It could be that they are all cheating: what about those wearing spectacles, contact lenses or hearing aids? We have to consider what allowances have to be made in special circumstances, for the competitive embrace increasingly includes the disabled as well as the able bodied. What about the special position of those other entrants who went to the Commonwealth Games – the disabled men and women battling for their medals at many of the same venues? Or the "Olympic Games" for those with heart, kidney or liver transplants.

Top competitors can take some medicines, provided they are approved. The problem is that the scheme that has been devised is full of inconsistencies. An elite athlete with asthma can take a beta stimulant, such as salbutamol (*e.g.* Ventolin) by inhalation but not, it appears, by mouth. However, with enough inhalations the amount reaching the blood is the same by either route. The same holds for inhaled corticosteroids such as beclomethasone (*e.g.* Becotide) – inhalation is acceptable, tablets are taboo. Salt tablets are permitted, but not tablets for getting rid of salt (*e.g.* diuretics). The oral contraceptive is rightly permitted, but there is no doubt it can help a woman with training and competition schedules as it allows for controlled and often shorter menstrual bleeding. Male (anabolic) steroids are forbidden although the unwanted effects listed for them (at least for males) are probably fewer than those for women taking oral contraception.

It may be argued that the sporting authorities have devised their scheme at least in part because they are concerned about the welfare of the athletes. But if that is so, why should they allow pain-killing injections into a joint which, although offering immediate relief, probably increase the long-term risk of arthritis? Alcohol and beta-blockers both steady the hand in precision events such as archery or pistol shooting. Both are effective, but only beta-blockers are on the list of banned substances. Should alcohol be, too? And why is cigarette smoking permitted? Smoking is probably the most dangerous drug practice in the world; it is certainly one that alters body chemistry, and it seems most probable that it affects athletic performance.

A parallel issue arises here. Should the sporting associations allow a sporting practice that, by its very nature, causes illness. The athletics world knows full well that excessive training, particularly in young women and girls, causes the bones to weaken. This is a very worrying problem but not one being tackled directly.

The issues involved in the control of "doping" in sport are not easy to resolve, but the present arrangements do sport a disservice. Their lack of consistency and their failure to tackle the medical and pharmacological issues make the present package untenable. The problem is that the schemes devised have been produced by the sporting associations in apparent isolation, and the implications are far wider than sport alone. There is a need for a group convened by some worldwide sporting authority, perhaps the International Olympic Association, to tackle these issues.

It is important not to leave policy making to bodies responsible for individual sports, or to national governing bodies. Preferably, the code of practice should cover all asports – professional and amateur – in all countries. Implementation alone would be a local matter. We have numerous codes at present and the principle is undermined every time there is an inconsistency.

The group, which should include physicians, pharmacists, lawyers, ethicisits, health specialists, and representatives of the sporting fraternity itself, must look for a scheme that is consistent, covers sport generally and can be simply implemented. It must also equip itself to deal with future problems. If a drug was found that could prevent the bone damage in those who over-train, or one that could reduce the rates of troublesome infection in top-class athletes, would these be usable? It would seem reasonable to offer every means possible to reduce illness and prevent future damage, but with current attitudes it is unlikely that such preventive measures would be permissable.

The present arrangements will not do, but whatever scheme is adopted it must be able to cope with at least any immediate developments. Perhaps the end-point must be the welfare of the competitor. Any practice that might lead to damage would be made unacceptable. Anabolic steroids, which are prescription-only medicines, could be banned because they are not to be used without a doctor's permission. However, they should not be banned because they might enhance performance (this measure might at least reduce the run on illicit trading in impure or defective products). In addition, any practice that was actually illegal could not be contemplated. The use of drugs of abuse, such as cocaine or amphetamines, or opiates for recreational purposes, should be outlawed. But their banishment should be because of their legal status rather than because they might give unfair advantage. The advisory group on drugs in sport will have a lot to consider, but the need for sensible deliberations by a multidisciplinary group is compelling.

THE CRANNOGS OF SCOTLAND

NICHOLAS DIXON

Nicholas Dixon, a Research Fellow in the Department of Archaeology at Edinburgh University, is Chairman of the Scottish Trust for Underwater Archaeology which was formed in 1988 to develop and promote the subject. His work at Oakbank Crannog, Loch Tay was the first underwater excavation of a submerged settlement site in Britain. The crannog reconstruction is unique in experimental archaeology.

What are crannogs?

Crannogs are ancient loch-dwellings which today survive primarily as submerged boulder mounds and islands topped by stands of trees. (Figure 1) These defensive homesteads figured prominently throughout Scotland's past as flourishing waterborne communities that lasted for centuries and came to play an important part in clan refuge and warfare. They were occupied from as early as the Neolithic period some 5,000 years ago until the seventeenth century AD, a testimony to the skills and ingenuity of early craftsmen and to the security provided by island living.

Most Scottish crannogs appear to have consisted of a single thatched roundhouse, deliberately built out in the water for protection from invaders. Evidence from underwater surveys and excavations shows that the crannogs were built as freestanding timber pile-dwellings in the lochs of woodland environments, and as circular or sub-circular stone buildings on manmade or modified natural rocky islands in more barren environments.

Underwater archaeology at Oakbank Crannog, Loch Tay

Eighteen crannogs are located in Loch Tay, Perthshire, where excavations have been carried out periodically since 1980. The work is carried out entirely underwater and the archaeologists dive (Figure 2) for up to eight hours a day in the shallow but bitterly cold water of the loch. The preservation of organic materials is spectacular at the late Bronze Age/early Iron Age site of '*Oakbank Crannog*'. (Figure 3) Surviving structural remains include the original pointed alder piles of the supporting platform, floor timbers (Figure 4), and hazel hurdles forming walls and partitions, as well as the posts which once provided a walkway to the shore.

The finds from the site paint an amazingly clear picture of the lifestyle of the crannog-dwellers. Wooden domestic utensils, finely woven cloth, beads, and even food and plant remains have all been well preserved. They kept cattle, sheep, goats, and pigs, and produced dairy products including butter, in one instance found still sticking to a wooden dish (Figure 5) probably only discarded because it had split apart.

Most crannogs are situated opposite good agricultural land, and the discovery of a wooden cultivation implement together with grain and pollen evidence indicates a local population of peaceful farmers. They grew a range of cereal crops including *Triticum Spelta*, an early form of wheat previously thought to have been imported by the Romans five hundred years later. The loch-dwellers also cultivated a taste for parsley which is not native to Scotland and indicates trade with people further south or on the Continent.

The crannog people supplemented their diet with a range of nuts and berries including hazelnuts (Figure 6), wild cherries and sloes, but they had to make an extra effort to pick cloudberries, which only grow high on the mountains. They also made special trips to collect branches of pine to make tapers or 'fir candles', the only pine found so far at Oakbank Crannog.

Laboratory analysis

Working underwater in the peaty waters of Loch Tay is difficult and cold but also very exciting bringing to light the remains of the house and objects left by crannog dwellers in the Iron Age. These artefacts have not seen the light of day since 500 years before even the Romans came to Scotland. But it must not be forgotten that the excavation is followed up by months of conservation and analysis in the laboratory.

Specialists examine samples of the deposits lying on the floors to identify the type of plants used and dropped in the house. In this way opium poppy and parsley from the Continent and cloud-berries from the high mountains have been identified. They analyse the microscopic pollen grains to give a clearer picture of the environment and how the people exploited it and they look at the insects still preserved in the site which give good indications of the conditions in the house. There are even parasite eggs still preserved in sheep droppings collected from where the animals were kept. While the massive timber piles and crosspieces and other structural remains are impressive the story is greatly enhanced by the discoveries of things that can only be seen through the microscope (Figure 7).

Reconstruction

While the lifestyle of the crannog-dwellers is becoming clearer, every discovery raises more questions. They can only be answered with further research. In an effort to address at least one unknown area a project has begun to show how our ancestors managed to build their timber houses in several metres of water. The richness of Oakbank Crannog has prompted many artists' impressions and a three-dimensional model (Figure 8) to show what a crannog would have looked like. Now, we are constructing a full-size crannog in the shallows of Loch Tay, overlooked by four of the eighteen ancient crannogs preserved there.

While it is a simple task to make a model with dowelling rods from the hardware store it is more difficult to work with full-size trees (Figure 9). It immediately becomes obvious that the crannog builders were strong and fit. Cutting piles to a point with a chainsaw takes only a few minutes but experiments were carried out first with a replica of a Bronze Age axe which takes much longer. While the modern builders have the advantage of modern tools they also have the problems of having to apply for planning permission and to conform to modern safety rules and regulations, clearly not problems confronted by people in the past.

While the planning authority demanded engineering drawings to be prepared to modern standards the engineers worked very closely with the archaeologists to come up with plans that were as close to the original site as possible. There are no joints or techniques in the reconstruction which could not have been produced by the ancient builders and there is not a single metal fastening in the whole structure. Often the modern workers would have been grateful for a handful of six-inch nails!

One of the puzzles from Oakbank Crannog was how the builders had driven piles, eight or nine metres long, into the lochbed. These massive timbers would require to be hit by a very heavy weight to have any effect on them and it is clear from the experiments on the new site that the piles were pulled upright (Figure 10) and then, using their own weight bearing on the sharply pointed end, they were screwed or twisted into the bottom. Hundreds of crosspieces, that hold the piles together and create a solid foundation to support the house, were carefully measured, cut and joined together with oak pegs (Figure 11).

The most common timber in the reconstruction is alder, as it was in the original, but many oak crosspieces have been used as well. The evidence from the site suggests that the early Iron Age people were managing their woodlands and ensuring that trees grew straight. Now it is difficult to find straight alder and oak trees in a group as they are not commercially viable timbers in Scotland. The modern builders had to search widely through the woodlands to collect sufficient straight trees for the reconstruction. It is clear evidence of the close relationship that people in the past had with their environment and how they valued and protected it.

The alder floor of the reconstruction looks identical to the remains of the floor from Oakbank Crannog. (Figure 12) The walls of hazel hurdles are also very similar and the task of cutting, collecting and weaving the hazel rods into wall panels cannot have changed much over the centuries. No doubt the original builders were more skillful craftsmen but they also had much better material to work with as hazel is not coppiced now in Scotland and, as with the alder trees, it took a lot of searching and cutting to collect sufficient straight rods.

The order in which materials are collected takes us closer to the way of life of the ancient people. Trees and hazel are cut in winter and

The Crannogs of Scotland

Figure 1 *This small island, at the west end of Loch Earn, is what the remains of crannogs look like in the landscape now. It was probably inhabited about 500 years ago. (Photo: Nicholas Dixon)*

Figure 2 *Underwater archaeologist excavating a large wooden bowl which had a timber upright driven through it when the site was rebuilt in the Iron Age. (Photo: Barrie Andrian)*

Figure 3 *Oakbank Crannog is situated at the end of the walkway in the loch and is completely submerged all year round. The land on the shore was farmed by the crannog dwellers. (Photo: Nicholas Dixon)*

Figure 4 *Major parts of a secondary floor are still well preserved and fallen uprights, which once supported the walls and roof of the house, can be seen lying across it. (Photo: Barrie Andrian)*

Figure 5 *The numbers on this wooden dish show where the butter was still sticking to the inside after two thousand five hundred years. (Photo: Nicholas Dixon)*

Figure 6 *The crannog people collected many natural foods from the surrounding environment, including hazelnuts and acorns. (Photo: Nicholas Dixon)*

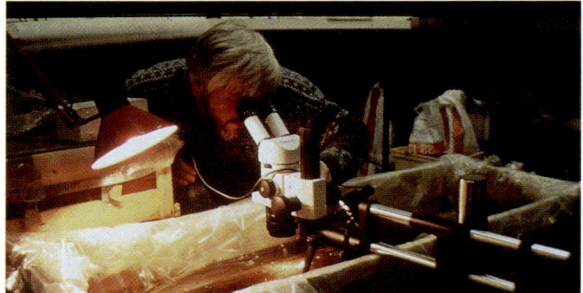

Figure 7 *The big structural remains from Oakbank Crannog are very exciting but some of the most important information is microscopic and is only recognised in the laboratory after the excavation is finished. (Photo: Barrie Andrian)*

The Crannogs of Scotland

Figure 8 *The three-dimensional structural remains at Oakbank Crannog supply good evidence for model building. (Photo: Nicholas Dixon)*

Figure 9 *Handling full-sized trees is very different from building models with twigs. (Photo: Barrie Andrian)*

Figure 10 *Two people are enough to pull very large piles into the vertical position for hand-driving into the lochbed. (Photo: Barrie Andrian)*

Figure 11 *Hundreds of crosspieces are accurately measured before being joined together with oak pegs to form the foundation of the crannog. (Photo: Barrie Andrian)*

Figure 12 *Compare this picture of the new crannog floor with the picture of the original Iron Age floor in Figure 4. (Photo: Barrie Andrian)*

Figure 13 *The crannog makes a comfortable and secure home and looks very impressive in the woodland landscape of Loch Tay. (Photo: Barrie Andrian)*

The Crannogs of Scotland

Figure 14 *Visitors to the Scottish Crannog Centre have great fun trying out ancient craft skills such as firemaking, wood-turning, nettle rope making and listening to Celtic stories and music played on the harp. (Photo: Barrie Andrian)*

spring before the sap starts to rise. This would give the builders the whole of the summer and autumn to build or repair the crannog. Recognising the correct type of tree and choosing correctly suitable trunks and branches for the construction is a skill that is acquired through practice and while the modern builders have only had those skills for a short time the ancient crannog people would have been familiar with wood and the skills of working it from their earliest childhood.

The most important part of the crannog reconstruction, and the part that suddenly gives it the appearance of a dwelling (Figure 13), is the roof. Hundreds of hazel rods are lashed to forty-nine alder roof poles resting on a ring beam at an angle of 50°. The steep angle is to allow rain to run off the water reeds that cover the roof so that it does not soak into them and rot them from the inside. Reeds are a superb thatching material and can last over fifty years in the right conditions. The crannog dwellers could also have used straw, bracken or heather to cover the roof.

The new crannog in Loch Tay is joined to the shore with a strong timber walkway which is based on the evidence from Oakbank Crannog although the new walkway is alder and the original was built substantially of elm. It is hoped that upon entering the house from the walkway visitors will feel that they are going back in time. The house will be divided into different functional areas; for sleeping, working and keeping the animals, with a clay hearth in the centre of the floor. There is no smoke-hole in the roof so there is no 'upstairs' in the crannog but it is possible that fish and meat were hung in the roof space to be smoked for the winter.

Building the crannog has given the archaeologists a clearer insight into the life of Iron Age people. They lived in a comfortable house with the security of knowing that they were protected by the loch from the robbers and bandits of the time. They had a healthy and varied diet of beef, mutton and pork and a wide variety of greens, vegetables, nuts and berries supplied by nature, and they surely caught salmon for which Loch Tay is still famous. They were skilled craftsmen working in stone, metal and wood and they lived in close harmony with their environment. It is no wonder that the crannogs were inhabited for hundreds of years and while the last crannog built in Loch Tay sits in a modern landscape it is hoped that visitors to the site will leave with a closer insight into the lives of the Iron Age crannog dwellers in Scotland (Figure 14).

The reconstruction, built by members of the Scottish Trust for Underwater Archaeology and Edinburgh University, is situated at the east end of the loch near Kenmore village and is the focal point for a new *Scottish Crannog Centre*. The project has been carried out with minimal funding, mostly voluntary labour and with a wealth of in-kind support from the local community and further afield. For more information and opening times, contact the Scottish Trust for Underwater Archaeology or the author at the Department of Archaeology, University of Edinburgh.

Peter Coveney is a senior scientist at the Schlumberger Cambridge Research Laboratory, United Kingdom, a Fellow of the Institute of Physics and a Fellow of the Royal Society of Chemistry. Before joining Schlumberger Cambridge Research, he was a lecturer in physical chemistry at the University of Wales, a junior research fellow at Oxford University, and a visiting fellow at Princeton University. He also currently holds a visiting fellowship at Wolfson College, Oxford University, in theoretical physics.

Dr Roger Highfield is the science editor of The Daily Telegraph. *He studied physical chemistry at Oxford University/Institut Laue Langevin and was the first to bounce a neutron off a soap bubble. He has coauthored two other books:* The Arrow of Time, *also with Peter Coveney, which was a bestseller. And* The Private Lives of Albert Einstein, *with Paul Carter. He has won several science writing awards, including a British Press Award, the Glaxo science writing prize, and a Commonwealth Media Award. He lives in Greenwich, London.*

LIFE AS IT COULD BE

**PETER COVENEY
ROGER HIGHFIELD**

Perverse, all monstrous, all prodigious things,
Abominable, unutterable, and worse
Than fables yet have feigned, or fear conceived,
Gorgons and Hydras, and Chimaeras dire
– **MILTON**

"Men will not be content to manufacture life: they will want to improve on it". With these words, the young Irish crystallographer John Bernal anticipated in 1929 the possibility of machines with a life-like ability to reproduce themselves.[1] He wrote of this "postbiological future" in *The World, the Flesh, and the Devil*: "To make life itself will be only a preliminary stage. The mere making of life would only be important if we intended to allow it to evolve of itself anew".

Almost two decades later, von Neumann performed the first work demonstrating the possibility of artificial life, in his efforts with self-reproducing automata. His so-called kinematic model aimed to isolate the logical content of biological self-replication. Though von Neumann conceived his self-replicating automaton some years before the structure of the genetic blueprint (DNA) had been unraveled, he laid stress on its ability to *evolve*. He had told the audience at his Hixon lecture that each instruction that the machine carried out was "roughly effecting the functions of a gene"; he went on to describe how errors in the automaton "can exhibit certain typical traits which appear in connection with mutation, lethally as a rule, but with a possibility of continuing reproduction with a modification of traits".

There were efforts to put flesh and bones on his ideas. In 1956, a proposal was outlined for "Artificial Living Plants", self-reproducing floating factories that could harvest important mineral and crop resources. Realizing the dangers posed if one such machine ran amok on the planet, Freeman Dyson of the Institute for Advanced Study in Princeton proposed a more benign *Gedankenexperinent* in which self-reproducing machines seeded life in the solar system. Today, however, von Neumann's realization of the logical nature of life has more significance than ever because of the ability of modern computers to evolve complexity.

We have already seen the trend to forsake rational design in biology and computer programming for techniques based on the blind forces of biological evolution. Imagine what happens when computer programs are evolved to solve hard optimization problems. Given that life is such a problem, is there any sense in which this might lead to new, computational forms of life? Indeed, is it possible to perform Darwinian evolution by "natural selection" the single property defining life inside a computer? While life on earth is restricted to carbon-based organisms, what we can create inside a computer is based on logical machine instructions; nevertheless, there is little to indicate that computer-based evolution does not have the potential for developing complexity on a par with that found in biology. Indeed, evolving machine codes should be able to generate *any* amount of complexity in that they may be capable of universal computation in Turing's sense.

Just as biological life ultimately emerges from the complex interactions of a great number of inanimate microscopic units called molecules, so some believe that artificial life (ALife) may emerge from complex logical interactions within a computer. The analogy between the two arises because all the logical processes that take place within a computer are based on the atomistic building blocks of Boole's binary symbolic algebra. The new science of ALife is predicated on von Neumann's abstract vision and is beneficial in several ways. As we will see, it can help us to better understand biological life by a more abstract study of the emergent properties that honed and shaped it, through the processes of reproduction, competition, and evolution. In turn, by harnessing within a computer the problem-solving capabilities of Darwinian-like systems, it will become possible to efficiently solve many nonbiological complex optimization tasks.

ALife provides a new way of coming to grips with the tricky concepts of intelligence and artificial intelligence by placing emphasis on the environmental influences on learning, as Turing foresaw long ago. Most intriguing of all, some believe that ALife may lead to the emergence of radically new forms of life. The first step along this path occurred some time ago with the birth of various computer viruses, and now more truly "living" organisms are coming to life within computers. At its heart, ALife aims to discover the essential nature and universal features of "life": not only life as we currently know it, but life *as it could be,* whether on earth, within computers, or elsewhere, and in whatever shape or form that it may be found or made within our universe.

Weak and strong artificial life

In a certain sense, artificial life research has been underway for decades, albeit by another name. Computers have been used to simulate a wide range of biological processes, which are often intractable by any other means. Typically, this use boils down to making a computer solve a specified set of equations that are believed to model the phenomenon in question, whether the pattern of limb growth or aggregation of a slime mold. Just as a computer model of the flash and mushroom cloud of a nuclear detonation is not itself a nuclear bomb exploding, so these simulations are in no way alive. Those who pursue such research can be called supporters of "weak" ALife, since they study computer models of biological processes in which the simulations would not be termed living. What distinguishes a certain strand of contemporary research in the field from the mainstream of study is an often unspoken belief in "strong ALife", according to which a suitably programmed computer may itself be deemed to be alive, or at least possesses properties of a living thing. This kind of distinction is quite familiar to researchers investigating artificial intelligence. There we find the "weak" artificial intelligence community, composed of people who use computers to model processes occurring within brains that could not be simulated without the power of the computer. The strong AI community, by comparison, has an even grander aim than the strong ALife group – to impute a conscious mind to a properly programmed computer. One way to distinguish these two AI positions is to translate the adjectives so that "weak" means modest and "strong" means daring. Today, AI is a sub-discipline of ALife, since only some forms of life can be expected to manifest intelligent behavior. But there is one important difference between AI and ALife as subjects for contemporary research: as the philosopher Elliot Sober of the University of Wisconsin has pointed out, terrestrial life is in many ways far better understood than the human mind, so there is a firmer grounding for ALife than for AI.

The term "life" is notoriously difficult to define. Since it is difficult to find two people who can agree on a definition of life, it might be thought that attempting to define artificial life would only increase the confusion. Petty arguments about definitions have not, however, discouraged the hubristic claims of the strong ALife community. One of the most provocative was made by Doyne Farmer of Los Alamos and Alletta Belin of Shute, Mihaly and Weinberger, in a manner reminiscent of the claims made by the proponents of strong AI. "Within fifty to a hundred years, a new class of organisms is likely to emerge. ... The advent of artificial life will be the most significant historical event since the emergence of human beings. The impact on humanity and the biosphere could be enormous, larger than the industrial revolution, nuclear weapons, or environmental pollution. We must take steps now to shape the emergence of artificial organisms; they have potential to be either the ugliest terrestrial disaster, or the most beautiful creation of humanity". It sounds outrageous. However, we need to examine the subject of ALife more closely before we glibly dismiss such views.'

Life as it could be

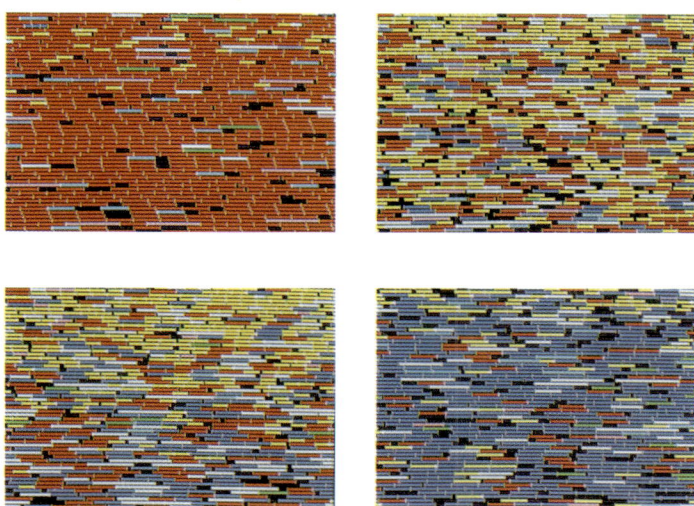

Figure 1 Evolutionary arms race between hosts and parasites in the *Tierra* Synthetic Life program. Images were made using the Artificial Life Monitor (ALmond) program developed by Marc Cygnus. Each creature is represented by a bar; the color corresponds to genome size (*e.g.*, red 80, yellow = 45, blue = 79). (1) Hosts, red, are very common. Parasites, yellow, have appeared but are still rare. (2) Parasites have become very common. Hosts immune to parasites, blue, have appeared. (3) Immune hosts now dominate memory, while parasites and susceptible hosts decline. The parasites will soon be driven to extinction. (4)

Figure 2 Fish and silicon chips. Computer-generated shoals display a host of realistic behaviors.

43

Viruses

Some examples of "weak" ALife forms are already in the environment and widely known. They were born during Core Wars, originally a game played by computer addicts. The idea was to create programs that compete against each other for processing time and space in computer memory-rather like animals competing for food and territory. Now there are unintended versions of Core Wars, better known as computer viruses, existing in PCs, workstations, and mainframes around the world.

The term "computer virus" is evocative; it was coined in 1983 for a short stretch of computer code that can copy itself into one or more larger "host" computer programs when it is activated. The threat posed by computer viruses is such that they are rarely out of the headlines and sport colorful names, such as Brain, Denzuk, Michelangelo, Elk Cloner, Festering Hate, and Cyberaids. They can also "mutate" when a malicious individual adds a minor adjustment to the viral code: for example, one of the most common, Jerusalem, has spawned Jerspain, Payday, Mendoza, Anarkia and Sunday, Fu Manchu, and Zero-time. The viruses have also become smarter. The middle of 1990 saw the first virus to use two separate strategies for replicating itself, either of which it could employ depending on the circumstances. Overall, in the time between the first infection by the Brain virus in January 1986 and April 1, 1991, two hundred different viral strains infected the IBM personal computer alone.

When these infected programs are carried out, the viral code is also executed. In the process, it may cause damage to the computer's operating system, overwriting important files, data, and instructions; moreover, the virus contains instructions for its own reproduction, thereby enabling it to spread via transfer to magnetic media such as disks and tapes. Other members of what is actually a menagerie of infectious codes also exist. For instance, "worms" are programs that can run independently and travel from computer to computer by the burgeoning global computer network, through the electronic equivalent of a chain letter. Other types of so-called vandalware include bacteria, Trojan horses, logic bombs, and trapdoors. Computer manufacturers have obligingly standardized the software in which they breed. Just as crop monocultures can be obliterated by a disease that would barely perturb a healthy mixed grassland, so cohorts of identical operating systems can be devastated by these seemingly insignificant computer infections.

Their proliferation begs an inevitable question: are computer viruses alive? This question is even more thorny than the same question addressed to bioviruses. The answer turns on one's definition of "life". Both natural and artificial viruses certainly possess some features of living things. They can reproduce themselves, they store information, and have a metabolism in the sense that they pirate the workings of a host, whether computer or cell. There are even examples of interactions between different computer virus species: the Denzuk virus will seek out and overwrite the Brain virus if both infect the same computer. However, one key difference can be highlighted with our understanding of hepatitis, a viral disease that has infected about half the world's population and is responsible for 90 percent of liver cancer.

There is not one "hepatitis virus" but many. Hepatitis B was the first discovered. Since then, A, C, D, and E have followed. Two new varieties have recently been suggested, and undoubtedly there are more. Surprisingly, these are not strains of one virus but arise from entirely different families, depending on their genetic makeup. But all are united in one respect. They are "solutions" evolved by quite different families of virus to the same problem: how to infect a human liver cell so as to reproduce. A and E enter the body in food and water contaminated with sewage; the other viruses exploit our enjoyment of sex and, like the AIDS virus, are blood-borne. This allows transmission from mother to child, via contaminated needles, and in such practices as ritual circumcision, tattooing, and ceremonial blood exchange.

Although scientists dislike anthropomorphizing what is little more than a complex molecule at the borderline of life, one of the pioneers of hepatitis research, Baruch Blumberg, admits that the strategies the virus adopts are so clever that it is difficult not to endow it with cunning intent. A virus consisting of a handful of genes seemingly plans endless strategies to outwit the human body thanks to its ability to evolve: the reproduction and mutation of viruses are so rapid that among every population a few are always able to adapt to a change in circumstances or exploit a new opportunity.

Unlike natural viruses, no man-made computer virus evolves; indeed, to create one that does would be a major programming challenge, given the intolerance-also called "brittleness" – of most computer languages to errors or "mutations". This is just as well, considering the damage that viruses can already inflict on computer systems. The most fruitful definition of a living system is one that is subjected to the rigors of Darwinian-style selection; at least by this definition such programs cannot be regarded as alive. Nevertheless, some strong ALifers, notably Farmer and Belin, still argue that although computer viruses need human beings to create them, many natural organisms cannot exist without the help of another. Thus, they argue, computer viruses have a symbiotic relationship with humans for their evolutionary development. Putting aside the sterility of terminology, it is clear that, although these artifacts comprise more or less disembodied information, they are disconcertingly close to being alive in the general sense.

Artificial growth

In some types of weak ALife research, computers have been used to model the processes of plant growth, a fiendishly complex enterprise. A few numbers can illustrate the problem posed by multicellular plants. Within each cell, thousands of genes are present. Not all are used within any given cell-the precise configuration of active and inactive genes depends on the type of cell, whether flower, stem, or seed, for instance. For n genes, the number of active combinations is 2^n. Even an unimaginably simple plant of ten genes would have 1,024 combinations of genomic states to spell out a cornucopia of plant designs.

Following in the footsteps of the British zoologist D'Arcy Thompson, in an attempt to capture the underlying universality of the genetic language used by all plants, the late Aristid Lindenmayer, a botanist at the University of Utrecht, developed a formal, mathematical description of plant growth. What grew from his interest over twenty-five years ago were intriguing simulations that are now dubbed L-systems in his honor. He tried to model plant growth using the same sorts of grammatical techniques that linguists use to analyze sentences. Instead of "parsing" plants as one parses a sentence so as to work back from real words to abstract parts of speech, L-grammars are typically run in reverse by starting with abstract "parts of plants" and using the grammar to guide repeated substitutions until one gets to components such as bark, stem, leaves, and flowers. The beauty is that there is nothing in the grammatical rules about the overall plant shape. The structure simply emerges from the computation.

Lindenmayer's computer models represent the body of a plant by a string of symbols, one symbol per module (for leaf, stem, and bud). The body "grows" as one repeatedly applies the algorithms for manipulating the symbols. Even simple systems of rules can generate "plants" that look real. Another step was to introduce branching to this bottom-up approach so that "tree"-like forms could be made. His models worked in a manner not unlike that of a cellular automaton. Most spectacularly, L-systems have been used to generate extraordinarily realistic plant forms similar to ferns. Two of his graduate students at the University of Utrecht in the Netherlands, Ben Hesper and Pauline Hogeweg, took the digital results from an L-System rule set and displayed them graphically on a monitor. In this

Life as it could be

Figure 3 *Artificial plants. A lilac twig and a field of flowers, generated by Lindenmayer's methods.*

way, they have produced realistic computer-generated images of ferns and of an aster, the plant that Lindenmayer was pictured holding at an Artificial Life workshop in Santa Fe in 1987. The image appeared in a dedication to his memory in a volume of papers from the second ALife gathering, held the year after Lindenmayer died.

Some scientists have used Lindenmayer's approach to see whether real plant mutants can be accounted for in terms of a change in specified developmental rules. Even film animation laboratories are now using L-systems to generate images of trees with computer graphics. His methods provide insights into morphogenetic processes, revealing how by the application of locally acting simple rules, whether genetic or algorithmic, branches, leaves, and flowers can evolve in parallel. However, L-systems are *not* designed to grow in the same way as biological cells. This is why they fall naturally within the broad field of "artificial" life studies.

Unnatural selection

Core Wars programs, computer viruses, and computer worms are capable of self-replicating, but not of evolving in an open-ended fashion. L-systems do not even have self-replicating capabilities. The ability to evolve is the central aspect of life as we know it, and attempts are now being made to do just this artificially, within cyberspace. For there to be any chance of creating genuine artificial life within a computer, we must find ways to introduce novelty by mutations and then select the "fittest" objects.

In September 1987 the first conference on Artificial Life was held at Los Alamos in New Mexico. The ALife gathering attracted 160 delegates, ranging from zoologists and plant biologists to chemists, physicists, and computer scientists, taking in along the way researchers working on automated Lego sets for children and others, all of whom shared a common interest in the simulation and synthesis of living systems, whether using chemistry, software, or hardware. The organizer, Chris Langton, reported that "the most fundamental idea to emerge at the workshop was the following: artificial systems which exhibit lifelike behaviors are worthy of investigation on their own rights, whether or not we think that the processes that they mimic have played a role in the development or mechanics of life as *we* know it to be. Such systems can help us expand our understanding of life as it *could* be".

Among those present was Richard Dawkins, the Oxford zoologist. He discussed how his *Blind Watchmaker* program could evolve "creatures" displayed on the screen of his Apple Macintosh personal computer. "Borrowing the word used by Desmond Morris for the animal-like shapes in his surrealistic paintings, I called them biomorphs", he explained. "My main objective in designing *Blind Watchmaker* was to reduce to the barest minimum the extent to which I designed biomorphs. I wanted as much as possible of the biology of biomorphs to emerge.

The biomorphs were generated by strings of computer code rather like bodies are generated by genes. The computer carried out minor changes ("mutations") in the code describing a biomorph, and

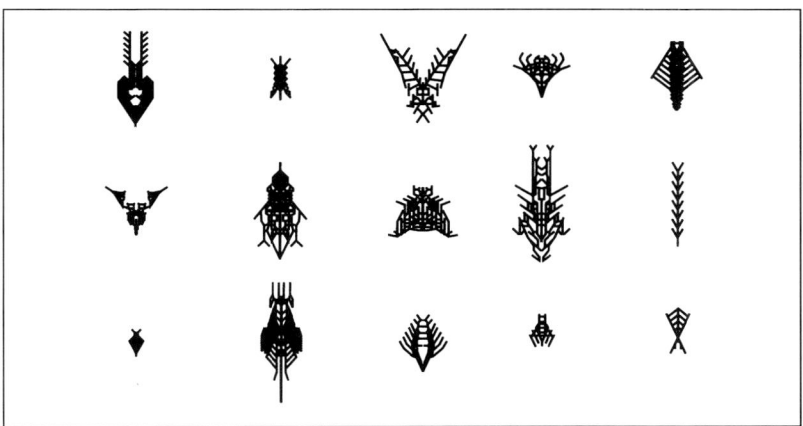

Figure 4 *Breeding diversity in a computer. Dawkins' biomorphs.*

displayed the range of body shapes that resulted. Being unable to select biomorphs according to how well they performed in an environment, Dawkins picked some for aesthetic reasons, and then bred new generations from them. After a few generations, surprisingly life-like biomorphs resulted. "I was genuinely astonished and delighted at the richness of morphological types that emerged before my eyes as I bred", he remarked.

Dawkins used his program to show how a "blind watchmaker" could produce the diversity of living things, without recourse to God or a grand designer. Dawkins, following in Darwin's footsteps, argued that the intricate design of the human eye could result from evolution by natural selection, through the interplay of chance and competition.

This was elegantly demonstrated in later work by Dan Nilsson and Susanne Pelger at Lund University in Sweden, who showed that if selection always favors an increase in the amount of visual information processed, a light-sensitive patch of tissue will gradually turn into a focused lens eye through continuous small improvements of design over a few hundred thousand generations. The study suggests that the evolution of something as complex as the eye could – in theory at least – have taken place in less than a million years, an eyeblink in terms of the vast span of geological time.

Evolutionary algorithms

Dawkins' biomorphs provide a striking illustration of how a series of random mutations can turn a simple structure into a life-like object. However, they are still a product of *unnatural* selection. It was a "god" – Dawkins – who provided the selection pressure that led to complex shapes appearing, rather than open-ended evolution through competition with other objects in the environment. As we have repeatedly stated, living things have an innate ability to evolve by natural selection, that is, via "survival of the fittest". Those species best optimized to perpetuate themselves in the complex but finite environment furnished by all other species and energy resources will be the ones that survive. Less than optimal creatures will eventually die out. Couched in these terms, evolution is rather like the many other hard optimization problems with the added complication that the fitness landscape (valleys and mountains) is itself coevolving, due to the individual struggle of all other species to survive. This means that evolution follows highly nonlinear dynamics, involving massive feedback loops, leading to a system that arguably represents the apotheosis of complexity.

Nevertheless, based on the sheer power of modern computers, we can be optimistic that the secrets of biological evolution may be simulated by computational processes. Recall John Holland's genetic algorithms (GA) which were inspired by Darwinian ideas. In the first twenty or so years following their creation, GAs have been largely used to solve complex problems within the inanimate world. However, even in the 1960s, Holland's group used GAs to investigate biological systems, starting with a simulation of single-celled organisms. Although initially slow to take off, the work using GAs, particularly in recent years, has provided a splendid artificial laboratory within which to dissect evolution. Genetic algorithms use an interacting population of digital codes, with each representing an individual organism, to model a population of organisms. The evolving codes "exhibit counterparts to such phenomena as symbiosis, parasitism, biological 'arms races', mimicry, niche formation and speciation". Other work with genetic algorithms has shed light on the conditions under which evolution will favor sexual or asexual reproduction, while Rick Riolo of the University of Michigan has observed genetic algorithms that display "latent learning", a phenomenon not dissimilar to learning in neural nerworks, in which an animal such as a rat explores a maze without reward and is subsequently able to find food placed in the maze much more quickly.

David Jefferson, working with an Artificial Life group at the University of California at Los Angeles, used a GA to develop a trailblazing "artificial ant", a digital insect existing within a computer, that consisted of a set of instructions inculcating it with a mission to "learn" how to follow a winding broken trail laid out on a grid, just as real ants follow the scent of a chemical messenger in nature. The ant was represented as a "finite-state automaton" – that is, as a string of binary digits that could only access a finite number of ant behaviors, or as an artificial neural network. The GA was used to evolve improved rules governing the next move of the ant once it had "sniffed" what was in the grid cell directly in front. The program started out with 65,536 (2 to the power 16) digital ants, an arbitrary number that suited the available computing power (one ant per processor). Within seventy generations on a massively parallel Connection Machine 2, it had evolved a significant population able to complete a twisting and tortuous trail of eighty-nine squares. This may sound like a lot of evolution but each generation "lived" for less than thirty seconds. In this way, the group demonstrated that it is feasible to produce, by evolutionary means, artificial organisms exhibiting complex, indeed life-like, behavior behavior that would be very difficult to design *ab initio* by writing down a computer program. With colleague Robert Collins, Jefferson went on to develop AntFarm, a computer program that simulates the evolution of foraging strategies in artificial ant colonies (see Figure 5a).

Genetic algorithms show how such complex behavior evolves. In return, the ants again illustrate how the use of many agents obeying simple rules can produce complex foraging behavior. Steve Appleby at the British Telecom Laboratories in Martlesham Heath has developed a similar ant program to simulate the qualities of ants that give them robustness their simplicity and ability to self-organize in the important job of gathering food. He hopes that these insects will inspire a solution to one of the most important problems facing any modern telecommunications company – how to route calls across a vastly complicated communications network.

Two extreme solutions are possible. In one, a supercomputer would sit like a spider at the center of this vast communications web. In theory the computer can always calculate the best way to route calls. However, in practice it may spend too long scratching its head and working out what to do next. At the other extreme, each telephone exchange could act independently, rerouting calls depending on how overloaded the neighboring exchanges are. This is fast, but never comes up with anything like the best solution because it lacks a global picture. British Telecom looked to the ants to provide a compromise between global and local control.

To persuade an ant colony to forage efficiently required just four rules. The first three exploited the way the ants communicate by leaving a trail of a chemical signaling molecule called a pheromone: (1) If the ant finds food, take it to the nest and mark the trail with pheromone; (2) If an ant crosses the trail and has no food, follow trail to food; (3) If the ant returns to the nest, deposit the food and wander back along trail. The final ingredient is a catch-all rule: if the other rules do not apply, wander at random (see Figure 5b).

"This pheromone trail is one of the keys to robustness", Appleby believes. "It is a message but is laid down for any old ant to stumble across. This is quite unlike the way computers function, where a program working on one computer sends a message to another specific computer. Robustness comes naturally as a result of not addressing messages. One ant can take over the job of any other that perishes. Working with Simon Steward, Appleby has now developed a swarm of "ants" – mobile programs – that can help route calls through a telephone network, evening out the load to ensure that each exchange is used efficiently and is not overloaded. Like the ants, these pieces of code do not communicate directly but leave messages for one another in each exchange as they pass through. "If one program crashes, it does not really matter because another one can come along later, pick up the message, and carry on. That is what gives the system robustness", said Appleby. BT is now exploring how to use such a system to manage dynamic networks, notably the networks that connect computers.

Open-ended evolution

Conventional genetic algorithms endow the quest for artificial life with some aspects of natural selection, but they suffer from numerous limitations. One is the sheer size of a simulation needed to reproduce the complexity of life as we know it. Natural populations can number millions in the case of large animals, trillions in the case of insects, and

Figure 5a *Ant farm. At the start of a new generation, all the ants are in the nest, there are no chemical signaling molecules – pheromones – and food is distributed around the farm. The virtual ants evolve foraging strategies in their search for food, laying down pheromone trails as they go.*

quintillions or more for bacteria. However, as massively parallel computers become more powerful, it should be feasible to evolve digital populations of increasingly realistic sizes.

Another limitation arises because the chromosomes that represent encoded solutions to a given problem the digital organisms are all the same size in a GA. Many of the most important parts of the evolutionary recipe have already been designed into the simulation,

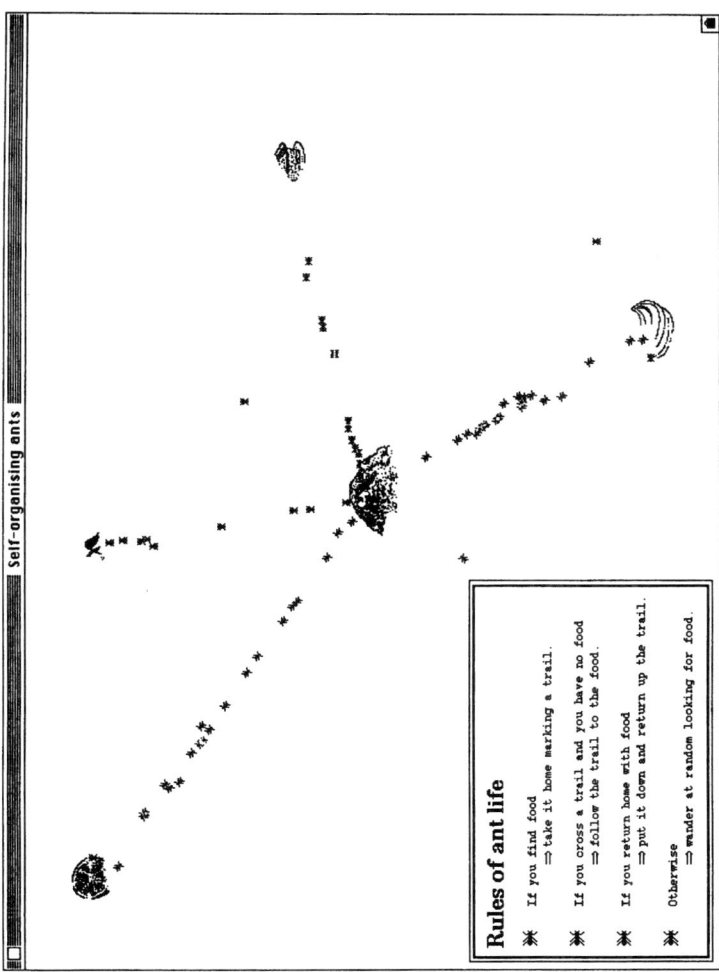

Figure 5b *Self-organizing ants, the inspiration for a new way of managing telecommunications networks.*

rather than evolving naturally. Moreover, the presence of "crossover" as a key genetic operator implies "sexual" evolution, while in nature the earliest and more primitive organisms reproduce asexually.

Furthermore, genetic algorithms employ "task-oriented" evolution while the real world uses "environment-oriented" evolution. In other words, an environment seething with life poses its own problems to be solved, rather than having one imposed on it. Imagine that the object is to breed a creature that can run at 40 mph. There are two possible approaches. Realizing that horses are already rather fleet of foot, you could launch a selective breeding program by taking a herd of horses, racing them, discarding the losers, and breeding only from the winners. This is the approach most current GAs take. Alternatively, you could select an open grassy plain like the Serengheti, populate it with lions, release a menagerie of animals, and then wait a few million years to see what evolved. It could be a horse, or a particularly nimble rhinoceros. However, all that matters is that the animal can consistently run at 40 mph. Very recently, it has become possible to simulate just this in GAs, so that evolutionary dynamics emerge spontaneously, just as they do in nature. Although less controlled, this approach is much more creative and some of its most enthusiastic advocates believe it may lead to genuine ALife.

Like biological life, genuine artificial life must be capable of evolving structures that are not designed or preprogrammed. To make the ALife evolutionary process open ended, it has to be random (stochastic); that is, it should embody an ever-present element of novelty that can cause change and evolution in quite unexpected ways. As the fitness of one organism alters, so the fitness landscape of its coevolving sister and brother organisms shifts accordingly. In this way, random simulations can produce solutions of a form never thought of or anticipated, imbuing a computer with originality and innovation. This aspect is essential for ensuring the plasticity of evolution-to prevent the whole process from running into a dead end.

The attempts at simulating evolution discussed so far have not been open ended; in other words, their "fitness functions" or landscapes are fixed once and for all. For instance, the effect of having genomes of fixed length in the GA is to tightly circumscribe the potential for innovation. In these models, a character string represents the organisms' genomes, which are mutated, recombined, selected, and replicated by unchanging design rules within the simulator. There is no mechanism of replication the genomes are just copied if they survive the selection phase. More must be done if full-fledged digital life is to emerge.

Virtual life: Tierra

Thomas Ray from the University of Delaware has created the first example of artificial evolution based on Darwinian principles. His achievement is likely to rank as one of the most important developments in twentieth-century theoretical evolutionary biology. Inside his computer, digital organisms fight for memory space in the central processing unit (the analogs of food and energy resources). Until Ray's work, we had not been able to experiment with evolution, only observe it. Merely observing it is far from satisfactory, since a single human life is as nothing compared with the time scale – "deep time" – which things happen in biological evolution. Ray's simulation of evolution, called *Tierra* after the Spanish for earth, provides a kind of metaphor for biological complexity, a tool to understand why it is seething with diversity.

Ray originally trained as a tropical ecologist and has spent much of his time studying vines in the rain forests of Costa Rica. He still lives there in his house-cum-laboratory set among forty acres of jungle. The inspiration for his computer simulation came not from the dizzying range of creatures that surround him, from humming birds to beetles, but the Chinese game of Go.

While a graduate student at Harvard, he once discussed the board game with a computer scientist, remarking on how the black and white stones used in the game form apparently self-replicating patterns, governed by the game's simple rules. The computer scientist made a fateful comment, saying it was equally possible to make a self-replicating computer program. That was new to Ray, who immediately imagined adding mutation and getting Darwinian evolution. He asked how this self-replication could be done and was told that it was "trivial". It took ten years before Ray was able to tackle this "trivial" problem.

Mutation and the ensuing competition for finite resources are crucial to evolution – it is not enough for an object to simply reproduce. That would lead to a "virus" with the banal ability to clog up a computer with identical progeny. Thus, for evolution to occur in *Tierra*, Ray not only had to create a self-replicating computer program, but also mutate it. Once this occurs, "natural" selection can take place: varieties of "organisms" best suited to their circumstances breed more effectively. The ensuing artificial life would then be a combination of self-sustaining complexity and reproduction, the latter allowing the former to evolve. It would also be evolution subject to the laws of logic, rather than physical laws.

There was, however, an important technical problem to overcome before this objective could be realized. Traditional von Neumann machine codes are not resistant to mutation: one period in the wrong place can cause a program to crash. This intolerance, the "brittleness" we have previously encountered, was originally seen as an almost insuperable problem by many ALife gurus. Farmer and Belin, for example, asserted that "Discovering how to make such self-replicating patterns more robust so that they evolve to increasingly complex states is probably the central problem in the study of artificial life". "I was concerned but not convinced", said Ray. "Why should a machine language have that property of brittleness and the generic language not?"

Inspired by a high-level programming language ("C") compiler and an accompanying debugging program that clearly revealed the inner workings of his newly acquired laptop computer, in 1987 Ray decided to construct his version of evolution using assembly language. Recall that this is a low-level machine-specific, binary computer code whose commands directly invoke the instruction set inside the computer's CPU, as well as services provided by its operating system. Since assembly language is so closely connected to a computer's hardware specification, it was the natural language for Ray to use for keeping control of the artificial environment within which he hoped to breed his digital organisms.

Ray made his machine code more robust with respect to random mutations (bit flipping) by borrowing ideas from molecular biology. The biological genetic code is characterized by a very small instruction set: there are sixty-four instructions (codons) formed from the nucleic acid bases, which get translated into twenty different amino acids. The *Tierran* language has thirty-two instructions all told, and thus is of the same order of magnitude as the genetic code itself. This represents a dramatic distinction from the situation that exists in conventional computers: even the new generation of RISC (reduced instruction set) machines have associated assembly languages comprising many more instructions.

Ray used three mechanisms for mutation, each reflecting parallel effects in biological evolution. Sometimes errors occur in the computations. Every now and then, a randomly selected memory location was changed, altering the binary code representing one digital creature; and whenever such an organism reproduced, there was a chance of introducing a copying error into its offspring. To avoid clogging memory, which would ultimately freeze the artificial ecosystem, a routine called the "reaper" kills the old and the error-prone organisms to compensate for the lack of direct predators. Deaths of organisms within *Tierra* occur because greedier or more successful organisms have monopolized the resources.

Ray also introduced "addressing by pattern" as a means of permitting direct interactions between organisms. This allowed one organism to exploit the instruction sets – the genomes – representing other digital organisms located nearby in computer memory. While no organism in *Tierra* could overwrite another's genome, it could read and execute that code. Ray describes the importance of such interactions in nature by using the rain forest as metaphor: in some parts of the Amazon, the physical environment consists of clean white sand, air, falling water, and sunshine. Embedded in that physical environment is the most complex ecosystem on earth, with hundreds of thousands, possibly millions of organisms. These do not represent hundreds of thousands of adaptations to the physical environment but hundreds of thousands of adaptations to other organisms: the organisms themselves become the dominant part of the environment, with the physical environment almost fading into insignificance in comparison.

The overall aim of Ray's work is clear: while life on earth is restricted to carbon-based organisms, the life we can create inside the computer is based on logical machine instructions; nevertheless, there is nothing to say that computer-based evolution does not have the potential for developing complexity on a par with the former. Indeed, the machine instructions within *Tierra* have been shown to be capable of universal computation, implying that evolving machine codes should be able to generate any amount of complexity, providing only that it is computable in Turing's sense.

In describing the *Tierra* simulator, it is helpful to keep in mind the analogy between Ray's version of artificial life and life on earth. Evolution on earth follows the basic principles of self-organization we outlined in earlier chapters. It takes place far from thermodynamic equilibrium, and is maintained there by the continuing presence of energy and matter. The energy ultimately derives from the Sun; the available matter is the material resource (food) that all living things must consume to remain alive. Under these conditions, matter becomes self-organized; the effects of self-replication and mutations, together with the fact that the resources are finite, means that a competitive situation develops.

Ray designed *Tierra as a* parallel computer with the MIMD architecture. He assigned a single processor that is, one with its own CPU – to each creature. Parallelism is only simulated on the serial machines Ray originally used, by assigning each virtual CPU time to run in turn (however, now there are versions that exploit *Tierra's* inherent parallelism).

Just as biological organisms use energy to organize matter, Ray's hope was that digital organisms in *Tierra* would use their time in the central processor to organize computer memory. A major difference between a more traditional GA program and *Tierra* is that while in the former the fitness function or landscape is controlled by the simulating code, in the latter it is defined locally by the individual creatures in relation to the environment. The fitness landscape thus evolves with the organisms.

Ray hoped that the initial self-replicating digital organism would evolve through mutations into distinct organisms that themselves might be capable of replication. In this way, the organisms would compete for the finite computer resources: CPU time and memory space. Then the creatures that survive should be those that evolve most effectively to exploit the others present. (See Figure 6.)

On January 3, 1990, Ray inoculated the virtual landscape in *Tierra* with life, starting with a self-replicating digital organism some eighty instructions long, *Tierra's* equivalent of a single-celled sexless organism. This creature, the first Ray ever dreamed up, was simply a stretch of instructions written in assembly code. It identified the beginning and end of itself, calculated its size, copied itself into a free region of memory of the same size, and then divided. In terms of the analogy with biology, the machine instructions of the self-replicating digital organisms would be better regarded as amino acids than nucleic acids, since they are "chemically active" – they manipulate bytes, CPU registers, the instruction set, and operating system. But the digital organisms Ray envisaged are perhaps closest to RNA-based molecular-biological creatures – which died out long before the Cambrian explosion 600 million years ago that led to a riot of complex multicellular creatures – because they bear the genetic information *and* carry out "metabolic" activity; in short, they represent both genotype and phenotype.

To discover if he had been successful in producing a cybernetic community, his computer displayed how many creatures of any given genome size were multiplying. All he could do now was wait. Not long after Ray started the simulation, he saw a mutant appear. Slightly smaller than the original, its population grew until it exceeded that of the ancestor. Other mutants needed fewer instructions still to reproduce effectively, and increasingly grazed in the available cyberspace.

Ray called a halt to *Tierra's* activities after several hours of processing time. It took a month-long follow-up study for Ray to tease out precisely what his digital organisms had been doing. *Tierra's* artificial society had rapidly become more complex. A creature had appeared with around half the original number of instructions, too

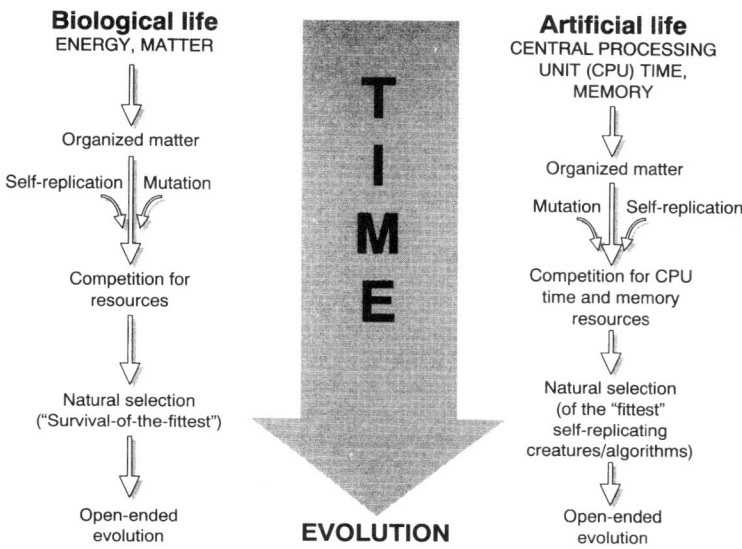

Figure 6 *The analogy between artificial and biogical life.*

few to reproduce in the conventional way. *45aaa* (as it became known) was dependent on others to reproduce. It was the first of many parasites. Later, *Tierra* developed hyperparasites – creatures that force other parasites to help them multiply, although they can reproduce in their own right. They were able to drive the previous generation of parasites to extinction by sharing key operating instructions. With the absence of parasites in the soup, "social" behavior emerged, when each creature relied on at least one other to reproduce. Gradually the community "evolved into a corner", producing a minimal organism twenty-two instructions long, compared with the original eighty. *Tierra* even showed signs of Eldridge and Gould's notion of punctuated equilibria, with lengthy dull episodes of elapsed time when not much happened (see Figure 1).

It is interesting to note that, once the original soup had evolved to a certain level of complexity, parasites emerged that did not replicate perfectly because of their genetic design. Indeed, these parasites played a role similar to sexual reproduction in nature in that they mixed up pieces of the genomes between different creatures at random. From this point, it seems that random mutations are not necessary to furnish the genetic change that drives *Tierran* evolution.

The result of all this was an effortless reconstruction of the kind of natural diversity that had seethed around Ray in Costa Rica. Just as the forest's rich diversity depends on "keystone predators", *Tierra* had keystone parasites without which artificial life fails. "I got all this ecological diversity on the very first shot", said Ray. "I was really excited – it was obvious I was getting my wildest fantasy. I thought it would take me five years of tinkering with parameters. The rich society of parasites adapted to each other rather than the environment in the computer. Once life starts, the evolutionary process feeds on itself "It is an autocatalytic, explosive process that builds its own structures so that the simplicity of the physical environment becomes irrelevant".

Although one might naively expect *Tierra* to favor the evolution of organisms that use less CPU time to replicate, which does indeed occur, much of *Tierran* evolution turns out to involve creatures discovering methods for exploiting one another. All of this emerges spontaneously – it is not preprogrammed – and is in full accord with the tenet of Darwinian evolutionary theory mentioned above: adaptation to the biotic environment (derived from the selection pressure provided by other creatures) rather than adaptation to the physical environment is the primary pressure driving the diversification of organisms.

Tierra was originally unveiled at the second Artificial Life workshop at the Santa Fe Institute, after Ray frantically completed the analysis of the results from *Tierra's* first successful run in his hotel room. His presentation was, unfortunately, "a complete flop", as he himself put it; he had to wait until he won a prize in an IBM supercomputing competition before his work got significant attention. Today that work is widely recognized as a watershed in ALife research. Ray makes the staggering conclusion that "It would appear... that it is rather easy to create life. Evidently, virtual life is out there, waiting for us to provide environments in which it may evolve".

There are, however, skeptics. Sir Robert May of Oxford University's Zoology Department, who has contributed much to modern theoretical biology through his work on nonlinear dynamics and chaos, finds the work stimulating but has "slight reservations about the extent to which the conclusions are perhaps inadvertently built into the program" as well as doubts about the robustness of the findings. Ray's response is to point out that "What we do in artificial life is create a universe, define the 'physics' of that universe and then set it in motion and observe what comes out. It is true that when you design the physics you predetermine what is possible in that universe". However, at the heart of these criticisms is the suspicion that Ray contrived the ancestral organism in *Tierra* to mutate in such a manner that communities would automatically follow. "That is definitely not what happened", he insists. "I am not clever enough to have 'built in' such a rich ecology and long chain of turns of the evolutionary race between hosts and parasites".

Plateau and C-Zoo
Despite the criticisms, Ray's work has been backed by other successful simulations of open-ended evolution. One was developed at Oxford University, inspired by Ray's success. Carlo Maley, then an American graduate student specializing in computer science and evolutionary theory, set out to test Ray's ideas for his master's degree at the Zoology Department. "I wanted to know whether it was hype or serious", said Maley. "It was important to me to write my own computer program so I could say something about whether Tom Ray had built things into his model or not, and whether it was a robust finding". He spent two years on one of the university's DEC workstations, attempting to reproduce Ray's work from scratch in a program called *Plateau*.

The name referred to the two-dimensional universe Maley created. While Ray used a one-dimensional string of programming instructions, Maley used a two-dimensional program so his creatures had shape: Loopy, which consisted of nested loops of instructions, was one of the two-dimensional creatures he set grazing on his computer memory. Unlike Ray, Maley found it difficult to get his model going because of bugs and because he lacked unpublished details of Ray's work. The two-dimensional programming language he used was also more brittle-less tolerant of mutations and creatures interacting with each other. But Maley did manage to iron out these wrinkles in his computer model. Then, within *Plateau, a* rich menagerie developed, albeit more slowly than in *Tierra*. "In any of these models, you will get some form of parasitism. That is the most robust thing about ecology in the real world and it was a surprising result that these models would show the same behavior", said Maley.

At British Telecom's laboratories in Martlesham Heath, another variant has been developed, based on a program called C-Zoo originally written by Jakob Skipper of the University of Copenhagen. Jose-Luis Fernández explained that BT selected C-Zoo because it had the qualities of *Tierra* but was easier to use. One problem with *Tierra* is that so many instructions in each organism are devoted to reproduction that it is difficult to persuade them to evolve to achieve another goal, such as finding their way through a maze. Another is that the results, displayed as a series of multicolored bars on a screen, are a headache to analyze. And the *Tierran* programs are hard to interpret because they evolve error tolerance through redundancy.

"The good thing about C-Zoo is that the biological ideas are the same as *Tierra* but it is much easier to understand", said Fernández.

In a typical C-Zoo demonstration, a series of ants scuttle about within a two-dimensional memory space, hunting for food, represented on screen by an apple. On the display an X marks a spot where there is more than one ant, and thus where a battle for memory space ensues. What we did was to substitute these pieces of computer code with a bit map of an ant. When you see an ant, what you really see is a collection of cells, each cell containing computing instructions", said Fernández. The code for each ant consists of only thirty-two different instructions, grouped as four cells of eight instructions. From these, complicated behaviors emerge, just as in nature complex proteins can be built from a set of only twenty amino acids.

Four instructions are all that we need to tell the ants how to move: move forward, right, left, or stop. The aim is to find food, either apples or other ants. Those that fail perish. Successful ants breed and mutate to produce progeny, with each different "species" being distinguished by a different color. When an ant dies it deposits its food – its code – into a central pool of memory where the other ants go. As a final touch, the screen display was toroidal so that an ant that wandered off the top of the screen reappeared at the bottom, and those that wandered off one side reappeared on the other. Fernández demonstrates self-organizing ant behavior by depositing all the food in the middle of the screen.

First, he starts with ants that could only move forward. Those that survived constantly return to the middle of the screen, where the food. Mutants that wandered off course would perish. After a while, green ants become the predominant species. Then he allows some ants to randomly mutate the instructions that control their movement. The ant army evolves to circle around the pile of food. "This feeding frenzy is an efficient solution and a very clear demonstration of evolutionary improvement", said Fernández. "The good thing about C-Zoo is that representation is much nicer compared with *Tierra,* which is difficult to analyze". As he discussed the program, a swarm of ants emerged that had their codes executed by nearby ants.

Virtual bees and fish

The computer simulations we have described so far are of virtual life, though no one would for one moment think that what can be seen on the screen of the computer bears more than a highly idealized resemblance to life on earth. Simulations are under development, however, that model real-world behaviors much more closely. At the

Salk Institute in La Jolla, Terry Sejnowski, Read Montague, and Peter Dayan have used a neural network to model and thereby understand how honeybees learn which flowers give them the best return for their efforts.

The focus of the biologists' interest is a brain cell called VUMmx1, which has connections extending throughout the bee's brain, according to studies by Martin Hammer of the Free University of Berlin. Some connections are stimulated by the bee's senses, as when the bee sips sugars, or detects aromas with its antennae. Others are connected to centers that control movement, or to the "mushroom body" responsible for learning. Through VUMmxl's connections, the bee learns to associate a stimulus with a reward, according to work by the bee psychologist Randolph Menzel, also of the Free University. What results is a memory linking flower scent to the nectar reward. Sejnowski reasoned that once these associations are set up, the bee can then use VUMmx1 to predict which flowers in a field are likely to yield the best rewards.

He then set about simulating a bee's brain. Though it has only about one million neurons, compared with our 100 billion, its tiny brain presents a formidable computing challenge: it can tackle about 10,000,000,000,000 floating point operations per second (10 teraflops), when today's most powerful computers usually work at less than one-tenth that speed. Sejnowski's neural network model contains enough detail to be biologically realistic, without oversimplification. "The model does not go down to the molecular level but it is not so completely general that it loses track of the individual neuron", he said.

The artificial neural network was trained on the basis of responses from "senses" to detect color and the taste of nectar. Depending on the timing of these responses, and a memory of the possible nectar reward, the artificial VUMmxl neuron predicted whether a particular flower was worth investigating, or whether to flit to another bloom. This network could be tested against the real thing because studies of foraging bumblebees had been conducted using artificial flowers by Leslie Real of Indiana University: the insects preferred blue flowers, which had been arranged to give a more consistent return of nectar, even when yellow flowers provided the same average return, albeit in an erratic way. Sejnowski's artificial bee had exactly the same preferences.

Another splendid virtual creature – a shoal of them to be precise – can be found in the Department of Computer Science at the University of Toronto. There, one can glimpse a striking "tank" of "virtual fish". The fish offer all the advantages of the marine world without any worries about feeding, cleaning out the tank, or disposing of the occasional victim of disease and aggression.

To enjoy the fish, all you need is a high-powered Silicon Graphics workstation worth a few tens of thousands of dollars and a copy of the "Artificial Fish World" program. The simulations were designed to emulate real fish as much as possible, capturing their form, motions, and behavior, by Demetri Terzopoulos, working with doctoral students Xiaoyuan Tu and Radek Grzeszczuk. The fish swim gracefully through water, scatter when pursued by a leopard shark, and compete for morsels of food. They even produce elaborate courtship displays. And yet they do not exist physically. Each is described by an individual computer program nested within a larger program, which generates a simple underwater ecosystem. 'We have demonstrated realistic-looking artificial fish that are capable of some astonishingly lifelike behaviors", said Terzopoulos (see Figure 2).

To write the program, the Canadians first used images of the real thing to give the fish coloration and texture. Next they gave the fish "brains", rules patterned after the real thing that control its twelve muscles, and "eyes" that enable them to perceive and react to their surroundings. The program took account of the mass and elastic properties of each fish, modeling it so that it is able to deform as it swims through simulated water. To coordinate the complex action of all the muscles, the fish learn to swim pretty much the same way a baby learns to walk", said Terzopoulos.

The fish tries random combinations of muscle actions, using an algorithm to refine their use. With the help of simulated annealing, the best combination is chosen depending on speed and, most important, swimming efficiency. After ninety annealing steps a virtual leopard shark hardly moves at all, because its muscles twitch randomly. After several thousands such steps, it can swim gracefully. "What comes out is very very natural", Terzopoulos said, "what ichthyologists call caudal locomotion because it depends mostly on the rear, caudal, fin". And, just as the undulating swimming motion is not programmed but emerges naturally, so the behavior emerges from simple rules: the researchers program each fish's affinity for darkness, coolness, and schooling, plus motivations such as its level of hunger, fear, or desire to mate. "Then we can start to develop predators and prey", explained Terzopoulos, "where the prey form schools, take evasive action, and scatter – as real fish do".

To model the elaborate courtship rituals found in the real world, the Toronto team studied the literature. "For some fish, the female displays an ascending behavior and the male goes underneath and nuzzles her belly", he said. "There are also courtship behaviors where the female and male circle around, chasing each other's tail". To create these displays, the team chained together such primitive behaviors as looping, ascending, nuzzling, in a sequence that depends on various events-for instance, the female has to witness the male's mating dance for a certain time before she responds.

Although the behavioral repertoire of the fish is programmed, what happens is highly complex and unpredictable because it depends on other fish in the neighborhood and what they are doing. "If the male gets interrupted by a predator, then the mating may not get consummated". Terzopoulos hopes to take the work forward by allowing the fish to mate by mixing the genetic components of male and female fish in forming offspring: "We may be within reach of computational models that can imitate the spawning behaviors of the female and the male, hence the evolution of new varieties of artificial fish through simulated sexual reproduction".

Animats

Another flourishing area of artificial-life research is that on *animats,* or physically real, yet artificial creatures. While the fish are virtual animats, efforts are underway to construct animat robots. Early attempts to build intelligent robots, in the late 1960s and early 1970s, centered around good old-fashioned AI (GOFAI). GOFAI is a top-down approach that compartmentalizes intelligence into discrete "modules" dealing with specific types of knowledge. One classic example was the "Shakey" project at the Stanford Research Institute, another "copy-demo" at MIT. The first robot navigated around obstacles in a room; the second piled blocks according to a model it was shown in highly structured environments.

In the new bottom-up approach to artificial intelligence, complex behavior emerges from the interaction of simple reflexes, making use of adaptive, learning algorithms that have been developed in the study of connectionism, that is, of artificial neural networks. The basic premise is that interesting robots are too complex to design. The pioneer of this approach in robotics is Rodney Brooks of the Massachusetts Institute of Technology, who inspired the field in 1989 with a prototype animat called Genghis that was based on a cockroach. Instead of being directed by a GOFAI computer program, Genghis had six independent legs that communicate with each other and operate by a few simple rules combined with an ability to learn on the job.

In general terms, Brooks' approach has two strands: a distributed, parallel-processing architecture, and the abandonment of the "compucentric" approach – explicit programming of a robot's brain with a preformed mental model of the world – in favor of real time control. Brooks wanted his machines to generate their own models of the world *adaptively* through their sensory experiences. "It is the coupling of the machine with the environment that is important", he maintains. The artificial nervous systems in these mobile robots is layered. As with the legs of Genghis, each layer contributes to behavior in its own right, although it may implicitly rely on others. For instance, an "explore" layer is not concerned with obstacles, because the existing "avoid" layer will take care of it. Building more and more sophisticated robots through progressive layering is analogous to the long-term results of evolution.

An animat clambering over an irregular terrain has no need for a prespecified computer program. "When the robot goes at different speeds, different gaits will automatically emerge", said Brooks. "It does not worry about which gait is the best or make an explicit decision". By adding senses, reflexes naturally emerge through adaptive control techniques. As one example, we can add a heat sensor so that if the animat senses more warmth on its right side, compared with its left, the range of the legs on one side can be altered so that it turns away. "The actual path the robot takes is not precomputed. It emerges".

Such is the speed and simplicity of the design that IS Robotics, a company set up by Brooks, has now commercialized Genghis with a Canadian company called Applied AI Systems. A swarm of twenty of these small robots was used to simulate the behavior of a termite colony. Large versions of the robots are being considered for trimming the undergrowth around trees in Canadian forests. A budget version of Genghis, called Marv, has been developed to negotiate rugged terrain of the kind encountered in places like Death Valley, California, in a project led by Chris Melhuish at the University of West of England, Bristol. The U.S. space agency NASA is also interested, in the wake of a 1989 paper written by Brooks for the *Journal of the British Interplanetary Society* entitled "Fast, Cheap and Out of Control: A Robot Invasion of the Solar System". It argued that rather than sending a single heavy-weight rover to explore a planet, a swarm of smaller robots would offer a cheaper, lighter, more robust alternative. NASA liked the idea of robots being fast and cheap, but it could not quite handle their being out of control, according to Brooks. NASA is currently considering a six-wheeled explorer for Mars.

There is a great deal of related work on animats. Neural networks that control robots are being evolved at the University of Sussex by Dave Cliff, Inman Harvey, Phil Husbands, Nick Jakobi, and Adrian

Thompson. Instead of designing fixed robot control programs, they rely on genetic algorithms to carry out simulated evolution on a random population of between sixty and one hundred dynamic recurrent neural networks. The networks control the behavior of simulated robots; their sensors and actuators closely model those of a real robot. During artificial evolution, selection pressure is controlled by evaluating the performance of each robot: the better the robot performs its task, the more offspring its cognitive architecture has. In one experiment, the Sussex team could evolve the networks to use visual sensors, "eyes", so that a palm-sized robot can avoid obstacles and seek light. Another wastepaper basket-sized robot used sonars, whiskers, and bumper bar-sensors to find its way around the Cognitive Science Department at Sussex University. "We apply selection pressure so that those that are better at doing the job we want them to do are more likely to breed", said Dave Cliff, whose colleagues have now turned to evolving control programs within a real robot. "We have some very promising results which suggest that we could evolve visually guided robots, which would find their way around a room", he said. "As far as we know, we are the first people in the world to do this".

Traditional AI researchers have scoffed that this kind of work has little to do with intelligence, because such animats do only what insects do. Brooks has responded with an ambitious plan to reproduce human evolution using a humanoid robot called Cog. Under construction since 1993, Cog has human form, with a head, arms, and even a voice, though it lacks legs (it was modeled on an MIT graduate student). "And we want it to have behavior like that of humans", he said. Sensing in the hands and arms is carried out with conducting rubber that allows touch. Strain gauges, heat sensors, and current sensors will allow Cog to feel its arms being used and how they are performing. Cog will also have two eyes, each of which consists of two tiny cameras, one giving a broad field of view, the other a central field. The eyes saccade-that is, dart back and forth like those of a human. Cog will have three microphone receivers for locating the origin of sound, a trick we manage with two ears and fancy signal processing in the brain. Such details are crucial, because the fact that we have bodies matters. Our bodies define, constrain, and enable us to interpret the world. The way the brain evolved and human cognition developed was predicated on our interaction as individuals with the world. "Any intelligence that we will really be able to communicate with better have a human-like body, otherwise it will be an alien", he said. "I did not want to build aliens but things that we can know and love".

Figure 7 *Some robots from Applied AI Systems.*

Cog sits in a public area in the AI laboratory so that it can interact with people, rather as a baby might, perhaps by playing with toys, stacking objects, passing things back and forth, and so on. Although Cog will have an ability to learn and its view of the world will be built through experience, it will still have some preprogrammed responses, just as a real infant is equipped with a sucking response and the tools for language acquisition. Cog has a sense of balance, and coordinates the movement of its head to keep the face of any nearby human in its field of view. "It wants people to pay attention to it", said Brooks. "That is its inner drive and we are hoping we can use that motivation for lots of developed behaviors". The information-processing power is located in an external computer that is connected via an umbilical cord. "My eight-year-old son was very disappointed when he found the brain was not in the head", he said. Understandably, Brooks is coy about when he expects Cog to evolve human-like behavior. "The six-month-to two-year-old level is the capability we are aiming at". He does not expect it to evolve a voice but only guttural responses. "We could have put a speech synthesizer on it but that is cheating", he said.

Evolving better programs

Clearly, the evolutionary approach to programming computers (and

Life as it could be

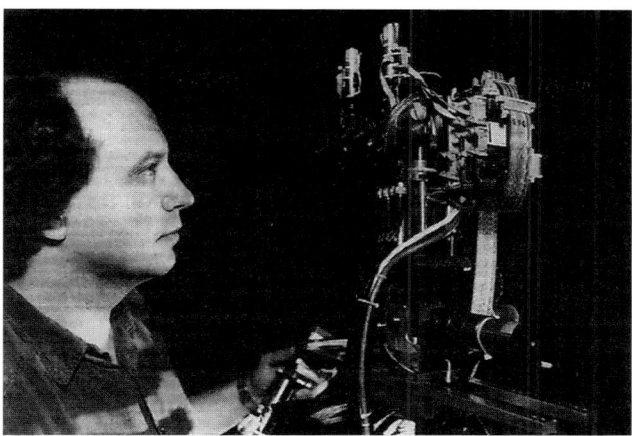

Figure 8 *Rodney Brooks and his humanoid robot, Cog.*

robots) is already dawning. Evolving communities of digital codes could be useful for the task of programming highly complex systems. British Telecom has already used the experience with *Tierra* and C-Zoo to calculate when we may expect to see people breeding software rather than writing it. Chris Winter, Paul McIlroy, and José-Luis Fernández at British Telecom started out by assuming that a good programmer can write up to thirty lines of debugged code every day. By their reckoning, a machine of an assumed processing speed of ten MIPS (million instructions per second), such as a present-day Mac Quadra, would take 100 days to achieve the same result. However, given the leaps and bounds in computing speed, they estimate that by the year 2000, desktop computers, able to crunch through about 3,000 MIPS, will generate computer code at approximately the same speed as a good human programmer. Indeed, it might even be possible to leave this software to breed in a computer so that it can evolve to cope with new challenges, such as infection with a computer virus.

Of course, tricky technical problems do exist. For instance, problems can arise because of shortcuts used in testing each generation of evolved software. It is computationally expensive to test every organism in every generation for a given task, say, routing calls through a telephone network. Instead a simpler test is used, measuring the organism's performance when one particular type of network traffic jam occurs. The danger is that the successful organisms are only good at solving that particular network problem and no other. One cunning way to overcome this problem is to evolve

69

the population of problems as well as solutions, so that the tasks we find most difficult are constantly selected. Developed by Danny Hillis at Thinking Machines, the method of evolving both problems and their solutions is called the "co-evolving parasite model" – a reference to the similar evolutionary race that occurs in nature between parasites and their hosts.

An illustration of how software may one day be evolved rather than expressly written down has been proposed by Tom Ray, as part of the ongoing development of his *Tierra* ALife simulator at the Evolutionary Systems Department of the Advanced Telecommunications Research Institute International in Kyoto. Ray wants to set up a "biodiversity preserve" by repeating the *Tierran* evolutionary computer experiments with volunteers on the International Network of Computer Systems, the Internet, which connects an estimated twenty-three million users worldwide. In collaboration with Kurt Thearling, he has developed software that would enable "digital wildlife" to multiply on the network. With the help of a few thousand volunteered computers linked across the world, the programs would be faced with a more complex environment that should encourage them to evolve yet more sophisticated strategies for survival and reproduction. Among the creations in this digital menagerie may lurk commercially useful software suited to spreading tasks among many different computer processors or doing jobs that Ray and Thearling have not even thought about. For example, it has been found that digital organisms can spontaneously evolve programming tricks, and the set of processes involved in *Tierran* evolution could be regarded as a general optimization technique for parallel programming. In Ray's words, "In the end, artificial evolution may prove to be the best method of programming massively parallel machines."

Because they will be chased off any given machine when its rightful user wants to get on-line, the organisms on the Internet will be constantly forced to try out new strategies, shrinking to the minimum size required for copying, learning quickly how to find little-used resources on the network. "They should start migrating around the globe, staying on the dark side of the planet", said Ray.

The creatures may start to cooperate, creating "multicellular" organisms. This adaptation of *Tierra* to a more collaborative multicellular form (as opposed to the "unicellular" one used to date) should lead to constructive cooperation between organisms in the solution of a single problem, rather than the fragmented effort on a variety of distinct ones typical in the unicellular case.

Various communication mechanisms are needed to coordinate composite organisms in *Tierra:* one operates like a hormone, sending a message from the mother cell to all its daughters; a second acts like a nerve, running between two specific cells. However, certain control tasks must only emanate from a single cell. For instance, when a mother cell divides to make a daughter cell, only the mother program should be executed if the pair need to take offensive action, or search for a resource or a mate. If mother and daughter cooperate successfully in this way, the first artificial multicellular *Tierran* creature will emerge. That could prove a milestone in the quest for artificial life and intelligence. As Ray puts it, "We are living examples of this kind of parallelism on an astronomical scale, with trillions of cells and hundreds of cell types that are beautifully coordinated. I am arguing that evolution has a proven ability to achieve that, just as I would argue that evolution is the only proven technique for generating intelligence".

Cellular automata and the edge of chaos
Various ALife groups have latched on to Ray's work, and today numerous people are working along similar lines. One variant on *Tierra* is *Avida,* developed by Chris Adami and C. Titus Brown working at Caltech.[65] While *Tierra's* organisms exist in a spatially nondimensional cyberspace, *Avida* is designed to introduce spatial dimensions into artificial evolution. It retains most features of *Tierra* but is based on a two-dimensional array of cells. The digital organisms occupy sites on the lattice, and in a manner characteristic of cellular automata they may only interact with their nearest neighbors. However, the update rules that determine how any site in the lattice is to alter from one time step to the next are not fixed once and for all prior to execution. Rather, the rules are determined by the genomes at the closest neighboring sites. These genomes change randomly via point-mutations of their strings of instructions, thus providing the impetus for evolution. As with *Tierra, Avida* can be used to perform tasks specified by a programmer. For example, if we wanted to evolve a method of performing multiplication of two integers, then we could reward every digital organism that accomplished this task by allowing it more slices of CPU time in which to execute. Over time, the artificial organisms will evolve to deal with this task. In principle, as with *Tierra,* such a system should be able to evolve to solve any task, no matter what its complexity, provided only that it be computable in Turing's sense.

Some scientists have argued that living things must be able to perform computations of essentially arbitrary complexity to survive the evolutionary arms race. Moreover, certain types of cellular automata (CA) are known to be capable of performing universal computation – in other words they can simulate a universal Turing machine. This claimed requirement for universal computation has inspired some people to revisit Wolfram's classification of CA dynamics, to see whether it furnishes any clues as to whether certain types of CA are better suited to artificial life.

As we have seen, Wolfram conjectured that his so-called Class IV CA would support universal computation. Lying between periodic and chaotic regimes, Class IV *appears* to support the most complex dynamics of CA. It might therefore seem reasonable to suppose that in ALife systems based on CA – in reality only a small subset of all such possible systems – the dynamics would have to evolve into Class IV when "life" arises.

Yin and Yang

These ideas on the relationship between dynamics and computation are part of a larger effort, notably in the field of statistical physics, to find complex behavior in "critical regimes" between order and deterministic chaos. The search for complexity in such regimes is also intellectually appealing since living things appear to capture an elusive mixture of yin and yang. Biological life seems to occupy a zone between regularity and turbulent chaos, where randomness coexists with creative adaptation. Organisms combine the ability to change and innovate with the stability of feedback systems that ensure a well-defined structure and metabolism.

Suggestive evidence for this kind of balance between chaos and order can be found within ant colonies. By using video recordings, Blaine Cole, working at Houston in Texas, was able to reveal that individual ants behave chaotically. For a while, an individual scuttles about. Then it takes a rest. And so on and so forth, in behavior that rattles around on a strange (chaotic) attractor. But at the colony level, the ants show quite rhythmic behavior. Nigel Franks, at the University of Bath, observed that the ant colony is active for a while, takes a rest, then becomes active again, with a cycle time of around twenty-five minutes.

Experimental studies by Cole showed that the pattern of behavior depends on the density of ants. If there are only a few in a territory, their behavior is chaotic. But if the number increases beyond a threshold value in a given area, the whole group becomes rhythmic. It is a bottom-up organizing process driven by ant-to-ant contact. The ants excite one another, so that an active ant gets an inactive one moving when they encounter one another. This behavior was simulated by Octavio Miramontes and Richard Solé, working with Brian Goodwin at the Open University. They developed a cellular automaton model of ant colonies in which the activity of individual ants is propelled by a Hopfield neural network.

"At a particular density of the computer ants there is a sharp transition in which what was a collection of individuals each doing its own chaotic thing suddenly transforms into a single whole-the colony becomes a superorganism with a well-defined rhythm and, at the same time, spatial order appears", according to Goodwin.

Observations from Nigel Franks' laboratory suggest that real ant colonies adjust their densities so that they live near this transition point, at the "edge of chaos". Ants regulate the size of the territory within which they make their nests, with the queen at the center and the developing ant embryos and larvae arranged around her. Given grains of sand, the ants define the boundary of the brood chamber. But if a malicious scientist pushes the sand grains in to cut the size of the chamber, the ants push them out again. Likewise, if the territory is increased, the ants reduce it again. "The colony has a sense of density and spatial order", said Goodwin. "So it appears that ants may indeed adjust their colony density such that they are near the edge of chaos".

These examples are what makes the "edge of chaos" idea so seductive. For Goodwin, it is "almost a theorem about life, the universe and everything that is complex and nonlinear (which is *nearly* everything). Speaking more anthropomorphically, the edge of chaos is a good place to be in a constantly changing world because from there you can always explore the patterns of order that are available and try them out for their appropriateness to the current situation. What you don't want to do is get stuck in *one* state of order, which is bound to become obsolete sooner or later (remember the dinosaurs, or the British Empire, or IBM before the shake-up). So complex systems that can evolve will always be near the edge of chaos, poised for that creative step into emergent novelty that is the essence of the evolutionary process. At least, that is the conjecture".

Evolution and universal computation

Chris Langton at the Santa Fe Institute, New Mexico, has been an ardent advocate of these suggestive ideas. He has devoted much time to an attempt to apprehend how life balances the yin of apparent mayhem with the yang of self-organization: "In living systems, a dynamics of information has gained control over the dynamics of energy, which determines the behaviour of most nonliving systems. How has this domestication of the brawn of energy to the will of information come to pass?" he asked. He claimed that cellular automata capable of performing "nontrivial computation" – including universal computation – are most likely to be found in the vicinity of transitions between order and chaos; that is the Class IV dynamics that lie "between" the ordered Class II behaviors and chaotic Class III behaviors.

Langton argued that if living systems perform complex computations in order to survive, evolution under natural selection would tend to favor systems near the border between ordered and chaotic behavior – at the "edge of chaos", where the ability to process information would be maximal. From a study of a large number of CA simulations using different dynamical rules, Langton maintained that the conditions for life are optimal when apparently random behavior coexists with more regular dynamics. Here "life" should be interpreted as either real or artificial.

Langton believed that he had found a critical transition region in the parameter space of all two- and one-dimensional cellular automata where "one observes a phase transition between highly ordered and highly disordered dynamics, analogous to the phase transition between the solid and fluid states of matter". This was the Class IV region between Class II and Class III. Phase transitions in physical systems, from ice to water or water to steam, are able to establish correlations between molecules over arbitrarily large distances in space and time. Since universal computation in Turing's sense can only work in systems with memory and communication over essentially arbitrarily large distances in space and in time, it appears that the complex (Class IV) category of CA dynamical states would be the most likely to support nontrivial and possibly even universal computation. Put another way, Langton was arguing that information processing within a parallel-processing network could be maximized at "the edge of chaos". One might then expect systems such as living things to evolve toward this region if they are to perform the many complex tasks their environment demands of them.

To bolster the "edge of chaos" concept, like-minded colleagues have tried to tie in other ideas. For instance, Stuart Kauffman has argued that the concept provides "a powerful new framework to understand evolutionary biology". In the natural ecosystem of a rain forest, a coevolutionary system depicted by a rubbery fitness landscape, the success of one species (such as a frog) may spell doom for another (a fly) that it prefers to dine on. Kauffman has claimed that the entire ecosystem may coevolve to a state poised at the edge of chaos, linking the idea with Bak's concept of self-organized criticality, who in turn proposed that a number of systems, including some cellular automata, exhibit evolution to a critical state.

One crucial issue remained: what is the *computational* capability that was supposed to blossom at the edge of chaos? This was addressed during the same period, but quite independently, by Jim Crutchfield and his colleagues at the University of California at Berkeley. Ctutchfield and Young showed that a system's computational capability increases dramatically at the onset of chaos, which is a kind of phase transition. Their work was published in 1989 at the culmination of a decade-long line of research involving Ditza Auerbach, Remo Badii, Peter Grassberger, Bernard Huberman, Gyorgii Sepfauluzy, Robert Shaw, and others.

Crutchfield's ideas concerned a neat method for describing the yin and yang of complexity in a statistical way. His definition of complexity led to a counterintuitive result. A string of ones represents pure order. A random string of ones and zeros would seem to be infinitely complex. However, Crutchfield and Young argued that, given a source of such random binary digits (*e.g.*, from a chaotic system), it is as easy to generate random bits strings as a string of zeros, so that utter randomness is as simple as pure order. Discounting the computational cost of randomness in this way, we can reveal higher levels of complexity between these two extremes: statistical complexity is maximized somewhere between order and randomness. In fact, Crutchfield and Young showed that there was a jump from finite to infinite memory at the onset of deterministic chaos. So in terms of their measure of complexity, computational capability is maximized in the regime intermediate between order and randomness.

For reasons that will become apparent, it is unfortunate that much more media attention has been focused on the efforts of Langton at Santa Fe and also the work of Norman Packard at the University of Illinois, who put forward the suggestion of the biological relevance of the edge of chaos and evolved cellular automata to perform a range of computational tasks. Using a genetic algorithm, Packard had investigated which cellular automata rules are "naturally" selected on the grounds of their computational efficiency and concluded that cellular automata rules associated with Class IV are the ones most likely to be capable of performing complex computations; he believed that if cellular automata rules are allowed to evolve so that they perform complex computations, those within this region would tend to be selected. Packard's work was promptly hailed-by a few scientists and the science media-as a landmark concept in the science of complexity.

Dissent at the edge of chaos

Criticisms of the work of Packard and Langton date back to at least 1988. "Serious flaws in its general reasoning and in the technical details were pointed out then and continue to be", commented Jim Crutchfield, whose important contributions have been overlooked in most popular accounts. Langton's work on the edge of chaos was qualitatively different from Crutchfield's in one respect. While Crutchfield and Young dealt with continuous time dynamical systems, the cellular automata studied by Langton exist only in discrete time. More important, Langton used a crude measure of complexity for his studies. The overall consequence, argues Crutchfield, "is that Langton's work cannot be construed as giving results on how evolutionary processes produce structure".

In a room opposite Langton's at the former site of the Santa Fe Institute, Melanie Mitchell also decided to investigate whether the media accounts of the "edge of chaos" had been inflated. She, too, was irritated by the vagueness of some of its claims. With student intern Peter Hraber along with Crutchfield, she tried to reproduce Norman Packard's important claim that his evolutionary CA simulations "adapted to the edge of chaos". They concluded that he was wrong. His "landmark" finding was almost certainly an artifact that said more about how his computer was programmed than anything else. In Mitchell's words, "To the extent that one can make sense of what Packard and Langton meant by the 'edge of chaos,' their interpretations of their simulation results are neither adequately supported nor are they correct on mathematical grounds".

In redoing Packard's computer experiments, Mitchell and her colleagues found no evidence for the idea that cellular automata with values close to the border between Class II and Class III dynamics had any particularly enhanced computational capabilities. Mitchell and her collaborators came to an important conclusion: "it is mathematically important to know that some CAs are in principle capable of universal computation. But we argue that this is by no means the most scientifically interesting property of CAs. More to the point, this property does not help scientists much in understanding the emergence of complexity in nature or in harnessing the computational capabilities of CAs to solve real problems".

High levels of computation at a *cellular automaton* phase transition between order and chaos have never been shown to exist. This topic is surrounded by confusion because of the suggestive yet ambivalent nature of the word "chaos". In everyday language chaos is synonymous with randomness, making people contrast it with ordered behavior, and thus think of some kind of precarious balance between opposites. But its scientific usage is quite different; there, as we have pointed out, the term masks the fact that chaotic dynamics is actually exquisitely organized.

Exploring complexity

In the final analysis, there is no simple nor lofty mathematical theory of life, whether real or artificial. However, we may draw strength from the ancient paradox that although the world is complex, the rules of nature are simple. The universe is populated by a rich variety of physical forms, from bacteria and rain forests to spiral galaxies, yet they are all generated and sculpted by the same underlying laws. Thanks to fast and powerful computers, biologists, physicists, and computer scientists pondering complexity can now explore complexity in its full glory, throwing light on questions that once lay exclusively in the province of philosophy and mysticism.

The most important contemporary ALife system, *Tierra*, has an evolutionary dynamic of such great complexity that the only way to investigate its behavior is to perform experiments on it within a computer, the equivalent of experimental biologists' field studies. One might anticipate that, given enough time and a sufficiently powerful supercomputer, it might be possible to evolve within a *Tierra*-like system digital organisms endowed with intelligent capabilities and even consciousness. The work of Brooks and others underlines that, for this to be possible, digital life must be exposed to and interact with the complexity of a rich and varied environment.

For us to pursue the quest for genuine artificial intelligence, we must now turn to the supreme manifestation of complexity in nature: the human brain. This is an object that fascinated von Neumann and Turing and provided the inspiration for much of their most significant work. Now more than ever before, it is setting the research agenda for scientists worldwide. The reason is easy to understand: it poses the ultimate challenge to the science of complexity, and to lay bare its secrets we will have to draw on many of the concepts and systems we have explored from mathematical logic and spin glasses to Belousov-Zhabotinski reactions and evolution.

THERAPIST

The shadow of a leaf falls upon the page
and as the breeze moves the shape skitters
    as if writing my notes
    Yes I am listening, really,
although what I hear matters much less
than what you say. You were supposing
    that you might be human
    complete with genitals, greed
and enough cowardice to survive a day –
just an example, not specially wrong.
    I agree (with this and more).
    Yet you imply your history
requires a unique and living repair.
I think that where you have come from
    depends on where you are now,
    but together we'll make a fable;
it seems we can manage a doubtful future
    provided we have a past.

Now you falter. Here we should attend
more losely. For though I am slowly learning
    those things of which
    it is better to be silent
for you perhaps the ache is too raw;
we'll honour your reticence, yet mark the place.
    There is time, oh indeed
    time will be your rack.
In the new quiet you gaze at me directly
and I half-expect you to surprise me, saying
    how you find in the end
    the personal is not enough.
I return your stare, and together we listen
to the gathering wind in the boughs outside.

Norman Kreitman

GIRLS (ONLY?) IN SCIENCE

MARTHA C. PHELPS-BORROWMAN

Martha Phelps-Borrowman has taught science to 11–14 year olds for 18 years. Currently, she teaches students enrolled in a gifted and talented program at Lanier Middle School in Houston, Texas. She is married and has a daughter and grandson.

Recently, I made one of my frequent trips to the main office of my large, urban middle school. I headed straight for a large array of drawers along one wall. Teachers' mailboxes occupy this space. I opened up my mailbox and papers jumped right out at me! You know, those teacher mailboxes that always seem to be stuffed with an endless accumulation of office memos, announcements, notes from parents, and myriad other items. As teachers, we add these items to our already tall stack of things to do. Now, I always go through everything because I know that every once in a while there is something really good hiding in there. This was one of those times.

A colleague had given me a copy of a publication by the Harvard-Smithsonian Center for Astrophysics. The booklet was titled "Space for Women." Along with it, my friend had included a special instruction for me to pay close attention to the back cover. I turned immediately to the back and read,

"Women hold up half the sky"
Ancient Chinese saying

This statement would likely evoke a variety of thoughts and feelings in anyone. But, my colleague had a sense of how I would feel about it. And she was right. I was deeply touched by the sentiment of that ancient Chinese saying. I am still impressed by it. It reinforces my pride in being a female. It gives me hope that young women will feel able to do whatever they want to do in life. This timeless notion would always have provoked some particular emotion in me. Its poignancy has been strengthened, however, along with my awareness of the place of women in the world – and especially the world of science and maths.

I have often looked back on the job of parent that I was blessed with and wonder what I did for my child. When my daughter, Stacy, was born, I felt very strongly that she could be whatever she wanted to be. Doctor, lawyer, Indian Chief… . I still have those same feelings, not just about my daughter, but about everyone. Over the years that strong feeling that I held for my own daughter's ability and future, didn't change. It was overshadowed, however, by the pressures of day-to-day living.

Most young women, Stacy included, need a lot of reinforcement to assure them that they can be whatever they want to be. Without parents' encouragement, the traditions set by society can come through loud and clear to both our young women and men. These traditions put women in particular roles. Some of the roles most often considered appropriate for women are mother, housekeeper, nurse, caregiver, and secretary. The roles of scientist and mathematician have traditionally been considered the domain of men. Even in the highly developed societies of the world today, the majority of our girls and boys still receive subtle and not-so-subtle messages of the traditional roles for boys and girls, women and men. These messages serve to close some doors of opportunity. Research indicates that this is especially true for adolescent girls and their pursuit of science and maths related careers.

Recently, as I began preparing for my presentation at the 8th Edinburgh International Science Festival, I became fascinated with a toy – a spinning top. It wasn't an ordinary spinning top. This one levitated! I started it spinning on a small piece of plexiglass about 20 centimetres square which was, in turn, placed on a special platform of the same size. I then raised the plexiglass and the spinning top to a height of 10–15 centimetres above the platform. When I removed the plexiglass, the top would hover above the platform for several minutes. What caused this top to levitate and defy gravity was simply the scientific principle of magnetism. This principle has always been in operation. But, as this gravity-defying top demonstrated, the application of the principle was somewhat different. This different application produced a different result.

I realised that I had been spending a great deal of time looking at an old idea. That old idea, like magnetism, has been in operation since humans began to inhabit the earth. It is gender-equity or rather gender-inequity, I should say – bias, favouritism, unfairness, or prejudice to someone because of that person's sex. These words do not bring to mind especially pleasant thoughts. Many people today don't consider themselves as being prejudiced. Yet, all forms of prejudice still exist. One form is gender bias.

The following paragraphs from *Roots of Gender Equity*, a booklet by Katherine Honey, were very effective in helping me to assess some of my own attitudes concerning gender equity.

A. In the scenario below, select the word that you feel best completes the sentences.

Ms. Blue, a manager of a chemical processing plant, interviews an applicant for assistant manager. She decides to hire one applicant because of **her/his** research background and experience in managing chemical engineers. Before Ms. Blue extends a job offer, she finds out that the applicant is getting married and wants to have children. Ms. Blue believes **women/men** who have young children are not career minded and that, ultimately, financial responsibility for the family rests with the **wife/**

husband. After considering the strong qualifications of the applicant and her own concerns for long term employment, Ms. Blue decides to hire another applicant.

B. Assume you have a son and a daughter. How would you complete the following?

When my children are grown my **son/daughter** may be outspoken and assertive and my **son/daughter** may be sensitive and nurturing. I would prefer my **son/daughter** to be a hairdresser or perhaps a teacher and my **son/daughter** to be a lawyer.

If I have to choose, I would like to see my **son/daughter** become the CEO of a large corporation and my **son/daughter** to be the primary care giver of my grandchildren.

When I read these paragraphs, I knew the choices that I **should** make. Yet, my mind still settled on the **traditional** choices. In the first paragraph, a man would be given the job of assistant manager, because a woman would be the one who would want to get married, have children and not be career minded. The husband would be the career-minded one and most people's choice as the primary breadwinner of the family. I was uncomfortable with the choices that I made in those paragraphs. I didn't think that I was biased or prejudiced against anyone. Nor did I feel that I held on to traditional stereotypes for men and women. However, I was surprised to find out that I have been gender biased.

Now, I can't tell you just **how** surprised that I really was. Me – biased – and against girls!!!

I just couldn't believe that I was unfair, especially to girls. I have a daughter, for goodness sake. I couldn't believe that I hadn't done all that I could to encourage her to do whatever she wanted to do. I began to look seriously at my parenting and my teaching and at the signals I was giving the girls and boys in my classes. The most frustrating discovery was that I didn't even know that my behaviour was perpetuating gender inequity.

How did this happen to me? Now, there's an old problem. I think it happened to me just like it probably happened to all of you, **and** your mothers and fathers, **and** their mothers and fathers, and so on back to the beginning of time. Just as it is happening now to our daughters and our sons. It is tradition. We live by traditions. The traditions of who and what girls and women should be **and** what boys and men should be are manifested all around us. We see it in all forms of media, our schools, our homes, and our workplaces.

Our young children are bombarded with messages about who and what they should be. Look at any number of girls' magazines for example. One that I purchased recently included such articles as, "love scope ... is there a guy in your future?", "Guy Shy? Get over it?", "Music Superstars – the bands ... the babes", and "Guy-Rated Looks." What messages do they give about how a girl today should look or think? Then, there are the movies and the messages that many of them give about how girls should look, or what girls should be doing. As long as these media are making money using this approach, they're not going to change a lot. The implication, is not that these books are "bad" in the information they offer young people. The point is that the picture these books offer must not be the **only** one available.

The sad fact is that, unless they are given other models, girls and boys, women and men are going to continue to see females as worthwhile only if they project a particular image. Much too often young girls, and older women alike, feel inferior because their appearance is unlike what is portrayed in the magazines, on television, and in the movies. This feeling of inferiority can lead to a loss of self-esteem and with it, a loss of self-confidence. This situation can further lead to women falling short of their potential.

I continue to be amazed at how ingrained the attitudes are about the roles for which women and men are suited. I thought we had gone far beyond seeing women as intellectually inferior. Or that only boys can play with trucks and only girls get baby dolls and play house. But, have we really? Although most mothers and fathers raise their children, both girls and boys, to have the confidence that they can do whatever they want to do in life, many parents still subtly forward the old roles for men and women. Society, in general, promotes these same attitudes, as do educators.

Our schools exhibit gender bias in several ways. Some of the major areas of bias are found in curricula, learning materials, the learning environment, and interactions between teacher and students and student and student in the classroom. Textbooks regularly present countless examples of stereotyping. Although there has been improvement, men and women are still frequently assigned traditional, rigid roles and attributes. Groups of people are often under-represented in the materials schools use with students. Textbooks frequently present only one side of an issue and continue to present material using principally masculine words and pronouns which serve to deny the positive contributions made by both sexes. These forms of bias can be counteracted by a parent or teacher who is well aware of the situation and chooses to do something about it. The unfortunate fact, however, is that most teachers and parents are not aware of the role they play in perpetuating inequity in education. Nor are they aware of some simple actions that can reverse the effects of inequity on their children.

When I looked closely at what I was doing as a teacher, I found that my behaviour might be contributing to gender bias for the girls and boys that I taught. I had a habit of calling on boys more than girls and engaging in more conversation with boys than with girls. I presented very little of the work of female scientists to the students. In fact, in all of the materials that I used, I had only one picture of a woman in a traditionally male role on display. I saw that I most often used male pronouns or proper names in most examples I gave to my students in the form of tests, questions, and scenarios. It was typical of me to ask the boys to do more "dirty" work in science than girls. And never did I carry out any sort of evaluation of curriculum materials for bias. Though you might think otherwise, research indicates my classroom behaviour is the norm. I then began to look at the old ways I'd done things and to engage in the discovery of new and different teaching practices. My objective was to produce different, more desirable results for the young people that I would teach.

We have gender-inequity in education which manifests itself in many problems. But, what do these old problems have to do with girls and science? There is little doubt that the problems created by gender inequity have created barriers to girls and their continuing interest in science, whether in school or in choosing careers. Their options are narrowed. Research is full of statistics on the factors that seem to turn girls off to science and mathematics. I read the statistics and the reasons with great interest. Then, to simplify everything that I had been reading and studying, I developed the following list for consideration.

GIRLS IN SCIENCE
Girls tend to like science until adolescence when their interest begins to diminish.
Girls do not take higher science and maths classes as often as boys.
Girls do not go into science and maths careers as often as boys.
Girls self-esteem decreases more than that of boys.

The more desirable results that I wanted to achieve in my attempts to remove the barriers to girls' pursuit of science were the opposite.

DESIRABLE RESULTS
More girls will have a sustained or increased level of interest in science.
More girls will take higher level science and maths classes.
More girls will pursue science and maths related careers.
More girls will develop a higher level of self-confidence.

What can be done differently? Where should the search begin for ways to be more gender equitable? I have found that one of the best ways to begin is with a self-assessment of one's own attitudes regarding gender equity. Simple assessments that are found in *Roots of Gender Equity* by Katherine Honey, as referred to previously, provide a good starting point. Recognising that there is a problem and that one may be contributing to it, must be the first step. Once that is accomplished, one is well on the way to creating an attitude of change.

In my classroom, my decision for action was based on the assessment of weaknesses that I had found in my teaching. I have made a conscientious and concentrated effort to call on the girls at least as often as boys. I focus on talking with the girls during class an amount of time equal to that spent in conversation with the boys. The walls of my classroom include more pictures of women engaged in science. Further, I have included pictures and written examples of women in "non-traditional" female roles. I make sure I include girls' and women's names in test questions I prepare, and in other written and oral scenarios in which names are used. I ask the girls to do any classroom task that I would ask a boy to do. I use many books and learning materials I have found which give biographies of women scientists and mathematicians from all around the world and amplify on the scientific endeavors in which each has been involved. The situations which would allow me to use materials which highlight women, or in any of the actions that I take, are not contrived. Students might easily see through such inventions. Instead, the addition of information in the natural flow of the studies in class is a preferred method for displaying the positive contributions of women to science and mathematics.

These actions seem quite simple. One might even argue that my previous classroom practices would have had little negative effect on the students and, likewise, the simple corrective actions would have little positive impact. Again, research indicates otherwise. As simple as these efforts are, I still have to make a conscious and concentrated effort to ensure that I carry out my plans. I can imagine how the traditional stereotypes might be projected by someone who does not even perceive the problem of gender inequity.

During the 1994–95 school year, a fellow teacher asked me to come visit two of his maths classes to conduct tallies of his classroom practices. He had me observe two specific teaching habits. The first involved the number of times he called on girls and the number of times he called on boys. He was extremely surprised when I gave him the results. He had called on boys almost 50% more than the girls – **and he knew what I was looking for**. He changed his practice drastically during the next class when I again conducted a tally on the same teaching behaviour. The second aspect was the number of times he called on boys versus girls when the girls had their hands raised. The results of this were surprising also. He still called on the boys more **even when the girls had their hands raised more often to answer**. These tallies served to make him aware of his habits and enabled him to equalise the numbers. This example illustrates how simply being conscious about the problem can produce a useful result.

My study of gender equity actually began when I was a resident teacher in the Rice Model Science Laboratory, a science project developed by Rice University in Houston, Texas. The project was conducted at Lanier Middle School, a public, inner-city middle school of almost 1,500 students. My female partner teacher and I were the first science teachers there to have an all-girl science class and, most likely, the first and only public school teachers in Houston to have an all-girl class of any kind. Because of the preparation that went into this unique teaching experience, my interest in how girls perform in science and maths, and in appropriate methods to sustain their interest in these subjects, grew measurably. I uncovered research on issues surrounding gender equity in education. I learned that there are a great number of individual educators and organisations the world over which have worked to heighten awareness concerning gender equity and develop special programs to work with girls in the areas of science and maths. It is significant that a larger number of parents and educators are aware of issues of gender equity in education. Without awareness, there can be no improvement.

My own self-improvement has been furthered by what I've done on a larger scale in teaching an all-girl science class, mentioned earlier, and developing an all-girl science club. At the beginning of the 1994–95 school year, while my partner teacher and I were waiting for the special all-girl class to be formed, we developed some assessment instruments to use with the students early in the term. We conducted the same assessment after the girls were in the all-girl class for the entire school year. Upon comparing the answers, we found some very surprising results. Here are the numbers. What would science be without numbers!

The first assessment required the girls to draw their view of a scientist. At the beginning of the school year, 67% of the girls drew a male scientist and 33% drew a female scientist. At the end of the year, the number drawing a scientist as a female nearly tripled to 89%, see Figure 1. A very pleasant surprise occurred when three of the girls drew their pictures of a scientist as young ladies in a science class. My own interpretation of this was that the girls were beginning to see themselves as a scientist.

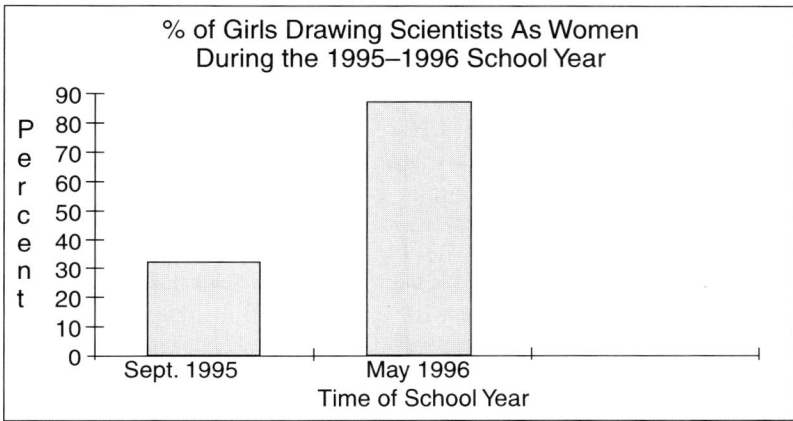

Figure 1 *A comparison of the percentage of girls in the all-girl science class who drew a scientist as a woman at the beginning of the school year and again at the ending of the same school year being a student in an all-girl science class with female teachers for the entire year.*

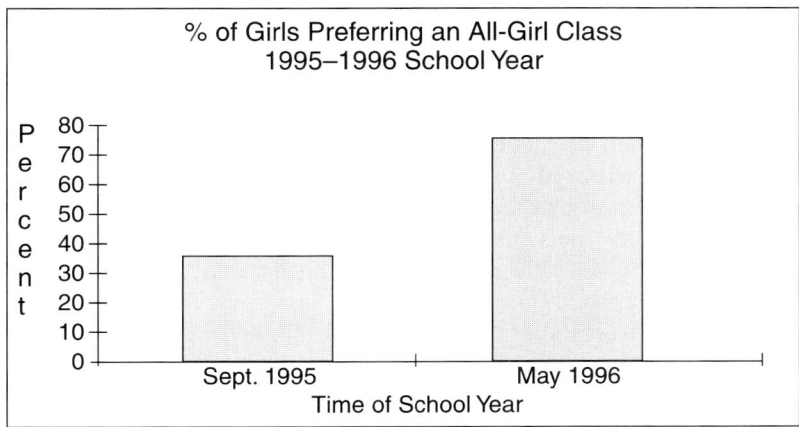

Figure 2 *A comparison of the percentage of girls in the all-girl science class who said they would like to be in a class of only girls at the beginning of the school year and again at the ending of the same school year after being a student in an all-girl science class with female teachers for the entire year.*

Another part of the assessment included a written attitude survey. One question was, "Would you like to be in a class of all girls?" At the beginning of the school year, 63% said they would prefer a class of both girls and boys. At the end of the year, the same group of girls responded with less than half (25%) preferring a class of both boys and girls, see Figure 2. This was a tremendous change in their attitudes about being in an all-girl class.

Most of the science experiences that were carried out with this class were identical to what would be done with boys. Some differences in the class were:

1. Guest speakers included several female scientists from the community.
2. All our time was spent in conversation with the girls. They had our undivided attention.
3. The girls did all of the work, even the messy, dirty things. They couldn't just sit back and let the boys do the experiments.
4. The girls cooperated with each other fully. This is often not the case in coed classes.
5. There was not the usual embarrassment the girls said they experienced when boys were around. They had more confidence in themselves. They weren't concerned about "not looking good" in front of the boys in class.
6. The distraction generated by boy-girl relationships was nonexistent.
7. For many of the girls, their previously held attitudes about themselves and science shifted.

On a local television broadcast, girls from the all-girl class the following year made some very provocative statements. One young lady said, "The science class is pretty cool because girls get to do what boys don't get to do ... because boys talk too much in class. And you don't have to worry about them staring at you and you staring back and telling them not to stare at you. With girls it's all the same. With boys it's a pretty nervous thing." Another of her classmates noted that, "Usually most girls are worried about how they look and what they're wearing when there's boys in the classroom. But, in here, they don't have to worry about boys. They just let everything out and say what they want to say. Everybody answers questions from the teacher. Nobody is real shy because everybody knows that they're not going to be made fun of (by the boys) if they get the wrong answer." These are individual viewpoints, but they mirror what most of the girls expressed in one way or another.

Are girls-only science and maths classes the best way to go? I'm not saying that they are **the** best way to go. I do believe that all-girl science and maths classes have been shown from research to be very effective in improving many girls' academic performance, especially in maths. They have also been found to increase the self-confidence of most of the girls, in their ability to do science and maths. I would certainly advocate them as an option. Laws in the United States permit single gender classes only if they are put in place to correct some discrimination in the past. There are many people who do not think that girls experience any discrimination in education and, as a result, there is always the possibility of a lawsuit for discrimination by not allowing boys into a class.

The second program that I was involved in was the establishment of an all-girl science club (G.I.R.L.S. – Girls Interested in Real Life Science) at my middle school. The club was funded by the American Association of University Women Educational Foundation. This foundation has carried out the majority of the groundbreaking research conducted in the United States concerning issues of gender equity in education and actions aimed at correcting the situation. It is one of many groups throughout the US working hard to ensure that all students have the same educational opportunities.

The goal of the girls' science club is to provide a mechanism for girls to come together in a supportive environment where they will interact with women scientists. This contact is designed to increase their interest in science and to foster the realisation that science and maths related careers are open to them. The club also includes elements designed to build the girls, self-esteem and confidence, such as the leadership opportunities that the club offers to the members through club officer positions and committee work.

The club has had great successes. There are also some circumstances that, while they are not to be considered failures, have been difficulties. The few that were encountered along the way, both in the all-girl class and the all-girl science club, arose principally from the attitudes toward gender equity held by teachers, parents, and students, both girls and boys. Some of the attitudes, in turn, were probably a reflection of how people may be threatened by having to deal with issues of gender equity. For others, the concept of gender bias in education created the need to face their own attitudes. More than a few simply did not believe gender equity was even a problem. In dealing with conflicting opinions held by others, I recognised very early the importance of avoiding a confrontational approach. I began to realise a feeling of achievement from knowing that even these adverse reactions brought a heightened awareness.

Another difficulty with the operation of the club, as with any large project, centered around time constraints. A teacher's workload is always heavy and the addition of another endeavour creates a further drain on any free time that one might have. The operation and logistics of the club required more adult assistance, support, and advice.

The G.I.R.L.S are so excited that I visited the Edinburgh International Science Festival to tell others about their club. They are proud of it. The overall success of the endeavour occurred because the girls actually took ownership. A few of the successes of the club are:
1. 50% of the members of the club are of ethnic minority.
2. Over 50% of the members attended a Saturday conference on science and maths for young girls.
3. Most of the girls have taken on some leadership role within the club.
4. The members have developed many ideas for operating the club.
5. Several projects initiated by members have already been carried out by them.
6. The club is continuing with the return of most of the original members who returned to the same school the next year.

Of the 19 girls who were in the all-girl class that I taught, only 15 returned to the same school the next year. Only 3 out of these 15 girls joined the girls' science club. Now, did the all-girl class or the girls' science club lead to them all saying, "I want to be a scientist when I grow up"? No, of course not. I do strongly believe, however, that some of the girls saw a possibility for themselves that they had not seen before, that of becoming a scientist or mathematician.

The presence of the girls' science club led to a lot of discussion among teachers, parents, and students, providing additional exposure to the issue of gender equity. Another great success is that the club has been replicated in other schools in the Houston area. Also, a successful group named Girls GetSMART (Science Maths And Related Technologies) was initiated in Edinburgh. The SMART project aims to encourage school girls ages 12–13 to take up Science subjects.

There are many simple things that parents can do with daughters to encourage them in science and maths. A few examples are:
1. Take your daughter to science festivals and events such as the Edinburgh International Science Festival.
2. Take your daughter to visit places like museums, science centres, and planetariums. Offer to take her friends and their parents, too.
3. Encourage your daughter and her friends to enroll in mathematics, science, and computer courses offered in summer university programs, or given by museums and science centres, and others.

4. Encourage her to volunteer at museums, hospitals, zoos, *etc.* during the school holidays or weekends.
5. Go camping or take outdoor family walks and hikes. During these walks identify plants, birds, or wildlife by referring to guidebooks for your area of the country.
6. Always show an interest in science and maths yourself. You could enroll in a summer class with your daughter at the local science museum or offer to volunteer to assist the instructor with set-up and clean-up of materials. Enroll in an evening science or maths class at a local university or community college. It just might be interesting and helpful to you while presenting a positive model for your daughter or son.
7. Avoid making comments such as "I was never good in maths either", "I didn't like maths at all", "Science and maths were very hard for me", or "I hated science and maths."

For an educator, there are a number of simple steps which could be utilised in the classroom to reduce gender inequity. Individual teachers could have a colleague come into the classroom to view a class in progress and conduct a tally of questioning techniques, or survey the room for materials (books, posters, worksheets, *etc.*) displayed which present females in non-traditional roles. This would offer a very good idea of present practices. Always be on the watch for examples of women in science and maths and use them in class when they enhance the studies. All of the simple things that I have done in changing my classroom practices are easy and effective.

Scientists in the world community can play a major role in improving the chances that our young girls will continue an interest in science and maths and perhaps enter into those careers. Some simple, easy practices include: visiting classrooms often to assist the teacher and students, offering to prepare and deliver a science or maths lesson, inviting groups or individual students and their parents to visit a laboratory or university classroom, offering to mentor students in science and maths and arranging for other scientists or mathematicians to do the same, and donating old equipment for the students to use in their science work.

Katherine Honey wrote in her booklet,

"Gender equity is not about girls and women beating boys and men. Gender equity is about people, regardless of gender, being given the same opportunities and expectations based on interests and aptitudes. A society that promotes gender equity values and rewards success the same for girls and women as it does for boys and men."

The attitudes and behaviours of society have really not been changed much over the centuries. So my message is, lets not try to change the attitudes of society as a whole. First, let's work to change the attitudes of our girls and boys, one at a time. They will bring about a bigger change in society. The cause is so important that every effort should be made by all members of society to ensure that each of our children live up to their full potential.

When I look back at the cover of that publication given to me by a fellow teacher, I know I am not alone. At least half the world is holding up the sky for our young girls.

For more information on gender equity in education go to this Internet site http://www.cs.rice.edu/~mborrow/GenderEquity/gendsite.html

Suggested reading list
Books
Schoolgirls: Young Women, Self-Esteem and the Confidence Gap, Peggy Orenstein. Doubleday.
Roots of Gender Equity, Katherine Honey. Katherine Honey & Daughters Inc., Publisher. P.O. Box 726, Attleboro, MA 02703 USA.
Science for Girls? edited by Alison Kelly. Open University Press, 242 Cherry Street, Philadelphia, PA 19106 USA.
Failing at Fairness, Myra Sadker and David Sadker. Touchstone/Simon & Schuster.
Options for Girls: A Door to the Future, An Anthology on Science and Math Education, A Project of the Foundation for Women's Resources and The LEADERSHIP TEXAS Alumnae Association in cooperation with Texas Woman's University.
Reviving Ophelia: Saving the Selves of Adolescent Girls, Mary Pipher. Grosset/Putnam.
The Courage to Raise Good Men, Olga Silverstein and Beth Rashbaum. Penguin.
Between Mothers and Sons, Evelyn S. Bassoff. Dutton.
Things Will Be Different for My Daughter, Mindy Bingham and Sandy Stryker. Penguin.
The Difference: Growing Up Female in America, Judy Mann. Warner Books.
How to Father a Successful Daughter, Nicky Marone. McGraw-Hill.
Women of Science: Righting the Record, edited by G. Kass-Simon and Patricia Farnes. Indiana University Press.
Breaking Barriers: The Feminist Revolutions from Susan B. Anthony to Margaret Sanger to Betty Friedan. Jules Archer.

How Schools Can Stop Shortchanging Girls and Boys: Gender-Equity Strategies, Kathryn A. Wheeler. Published by the Center for Research on Women, Wellesley College.
Beyond Dolls & Guns: 101 Ways to Help Children Avoid Gender Bias., S. Crawford. Portsmouth: Heinemann, 1996.
"Diversity in the Classroom: A Checklist," in *Common Bonds: Anti-Bias Teaching in a Diverse Society*, K. Matsumoto-Grah. Wheaton: Association for Childhood Education International, 1992.
Reflecting Diversity: Multicultural Guidelines for Educational Publishing Professionals. Macmillan/McGraw-Hill, 1993.

Publications available from American Association of University Women.
1-800-225-9998.
Sex Equity Handbook for Schools, D. Sadker and M. Sadker.
Growing Smart: What's Working for Girls in School, AAUW.
Girls in the Middle: Working to Succeed in School, AAUW.
How Schools Shortchange Girls, AAUW.
Hostile Hallways, AAUW.
Shortchanging Girls, Shortchanging America, AAUW.
The AAUW Report: How Schools Shortchange Girls, AAUW.
Ten Tips to Build Gender-Fair Schools, AAUW.
AAUW Issue Briefs. Set of five briefs on gender equity. AAUW.
What Parents Can Do, The Maths and Science Network for Expanding Your Horizons, AAUW.

MUSEUM

my niece is dancing round the museum
switching on steam engines and locomotives
beam engines and turbines
and the miniature huge pistons of extinct ships

child of the New World
five foot ten, fourteen years old and Canadian
she *just loves* the iron wheels
gears and shining cylinders
which rumble at her touch

the pride of tremendous bearded Victorian men
who kept their women at home in corsets

tall and pretty, in pastel sportswear and pink shoes
my niece dances from glass case to glass case
her finger gleefully hitting the red start button
first time, every time

neat, she says

Mary McCann

David Faulkner is a Fellow of St John's College, Oxford and a Senior Research Associate at the Oxford Centre for Criminological Research. He writes and lectures on various aspects of criminal justice and public administration, and is a Trustee of various charities concerned with the prevention of crime and the treatment of offenders. He served in the Home Office between 1959 and 1992, becoming Deputy Secretary in charge of the Criminal and Research and Statistics Department in 1982 and Principal Establishment Officer in 1990.

BASING POLICY ON EVIDENCE AND PRINCIPLE: HOW TO INFLUENCE DECISION-MAKERS

DAVID FAULKNER

It is one of the disappointments of the end of the twentieth century that, at least so far as the social sciences are concerned, the processes of government and the disciplines of science have moved so far apart. So too have the political and academic communities, with civil servants, other professionals and practitioners placed awkwardly between them. The increasing distance coincides with changing views of the state and of citizenship, and of the nature of public duty and the public interest.

This paper is concerned principally with the problems of crime and criminal justice and the services which are engaged with them, but the context is one which affects public decision-making, social policy and public service more generally. It is written from a perspective which is based mainly on experience in Great Britain, and within Great Britain in England and Wales, but some of the analysis may be recognisable in other English-speaking and European countries – for example, in Australia (Pusey, 1991).

This paper starts with an account of the largely evidence-based policies which were pursued during the 1980s, and the abrupt change of direction which took place in 1993. It offers some reflections on the social and political context in which public decision-making now takes place, the attitudes and assumptions which inform that process, and their implications for the management of change and complexity which is an inevitable feature of modern life in the criminal justice system and the public service as a whole. It sets out some possible parameters for an evidence-based and principled programme for crime and criminal justice, the form which such a programme might take, and the questions which would have to be answered in putting such a programme into effect. It finally considers the influences on the decision-making process and the means of effecting change.

An empirical approach to policy

The beginning of a scientific or empirical approach to crime and criminal justice – or, as it would then have been said, the treatment of offenders – can in Great Britain be traced to the late 1950s and early 1960s. It was associated with the formation of the Home Office Research Unit, the Cambridge Institute of Criminology, and the Oxford Centre for Criminological Research, and with other important work being developed in London and elsewhere. Prominent names included R.A. Butler (Home Secretary), Charles Cunningham (Permanent Secretary at the Home Office, previously at the Scottish Office), Leon Radzinowicz (in Cambridge), Nigel Walker (in Oxford, also previously at the Scottish Office) and Herman Mannheim (in London). The study of criminology developed quite rapidly from these beginnings. It is now well established in many universities in the UK and in most countries in the developed world, and there is increasing interest in countries that are in transition from former communist regimes. The influence of criminology on policy and professional practice has also increased, although progress in this respect has not always been straightforward or consistent, as the focus of political interest has shifted and the interpretation of the data has been challenged and modified.

New academic insights into the nature and extent of crime, the characteristics of criminality and patterns of criminal careers coincided in the 1980s with increasing interest in an evidence-based approach both among practitioners and in government. While researchers worked on evidence from the British Crime Survey and the Offenders' Index, and on subjects such as juvenile delinquency, crime prevention, the effectiveness of sentencing and the management of prisons, the police, prison and probation services were facing their own problems of identity and effectiveness – disorder on the streets and accusations of malpractice for the police, disturbances and industrial action for the prison service, loss of business and problems of credibility for the probation service. The courts' sentencing practice was shown to be inconsistent, and appeared to have little effect on the general level of crime. The government's Financial Management Initiative was demanding increased effectiveness and value for money. Both the Government and the services themselves were looking for a clearer and more rational sense of direction and purpose, based on evidence and practical experience.

The 1980s and early 1990s was, therefore, a period that saw a relatively scientific approach to crime and criminal justice, and a serious effort to take the evidence of research and statistical analysis into account. There was some disappointment that the material available gave inconclusive answers to questions about the causes of crime or the means of preventing re-offending, and some criticism from the political right that it did not address the case for an authoritarian or predominantly punitive approach. But there was enough evidence to show that attempts to reduce crime could not rely exclusively on the processes of detection, prosecution, conviction and sentence; that there were serious questions about the effectiveness of some aspects of policing, especially patrol (Clarke and Hough, 1984; Audit Commission, 1996); and that an indefinite expansion of prison accommodation, or of the prison population, did not provide value for money in terms of its effect on the general level of crime or on the subsequent behaviour of individual offenders (Home Office, 1990).

The result was the development of a set of policies which could be seen as having as their objectives to:

- prevent and reduce crime;
- establish a more coherent and principled basis for sentencing and parole;
- avoid the unnecessary use of imprisonment and stabilise the prison population;
- develop schemes of supervision in the community which would effectively reduce re-offending;
- give greater consideration to victims;
- make the system more efficient, effective and accountable.

(Windlesham, 1993).

The principles underlying these objectives included proportionality in sentencing; structured decision-making; due process and accountability in administration, including recognised and accessible channels for complaint; equity in the provision of services; and respect for individuals, including those who are members of ethnic or cultural minorities. They also included an empirical, or evidence-based, approach to the formulation of policy and the development of operational practice.

A change of direction

A shift of political mood took place at about the end of 1992, accompanied by a dramatic change in the Government's own political direction (Windlesham, 1996). The change was partly a response to genuine public concern about the continued rise in crime – a concern which was exploited, and perhaps fed, by intensive, sensational and often ill-informed reporting in the media. But its origin can also be traced to the internal difficulties of the Conservative Party at the time of the country's withdrawal from the Exchange Rate Mechanism, the controversy associated with the Maastricht Treaty and the need to find a popular issue on which the party could be united. Once the change was made, the other political parties felt they had to follow. At the same time, the practical managerialism of the early 1980s was giving way to a more assertive and dogmatic form of central direction, across the whole of government, based on notions of contracts and markets. Power was increasingly concentrated at the centre; and responsibility, and blame when things went wrong, was devolved to local managers. A new commercial model was being superimposed upon the professional, bureaucratic and (in the police and prison services) command structures which had competed with one another in the past (Freedland, 1994; Jenkins, 1995; Police Foundation/Policy Studies Institute, 1996). The evidence which counted became the analysis of costs and key performance indicators, and the principles became those of accountancy and financial management. Management consultancy was preferred to criminological research.

Attempts to analyse and understand social problems, and to resolve or prevent them from occurring in the first place, have given way to an obsession with punishment and social exclusion – for abuse of social security, for industrial action in public services, for disruptive behaviour in schools or by children more generally, for inconsiderate neighbours. Examples in the area of criminal justice include the Crime (Sentences) Bill, and the White Paper "Protecting the Public" which preceded it (Home Office, 1996a), with their array of proposals for "honesty" in sentencing and for mandatory sentences which would remove the courts' discretion in an important range of sentencing decisions. Others are the proposals for identity cards, access to criminal records, and the "shaming" and electronic monitoring of offenders, especially children. All are intended to generate a sense that they will "protect" the public from a dangerous, criminal underclass which is beyond reform or redemption and which will continue to prey on people "like ourselves" unless they are coerced or excluded from normal membership of society or the community. All the measures are based on personal conviction, political judgement and simplistic appeals to "common sense". Popularity with the public is argued as if no further justification were required. The proposals are rarely founded on any systematic or rigorous analysis of the problem or appreciation of the broader or longer term social consequences, or even in some instances (as in the Crime (Sentences) Bill) of the financial cost. (Penal Affairs Consortium, 1996; Hood and Shute, 1996). Green and White Papers rarely contain any reference to, still less any discussion of, the empirical evidence even when it is readily available. Professor David Garland has drawn attention to the weaknesses and limitations of this approach, and the problems of power and authority that lie behind it (Garland, 1996).

The social and political context

The movement towards policies and practices which respond and appeal to – but also help to generate – emotions of anxiety and fear can be seen as a natural but dangerous reaction to a situation of increasing uncertainty and declining confidence. That situation can be traced back over a period of many years. Nationally, the UK has for some time been unsure of its economic prospects, its place in the world and especially in Europe, the standing and even legitimacy of traditional sources of authority, and its long-term social coherence and stability. At the same time, individuals have been concerned about the prospects for their livelihoods, families and relationships, and sometimes for their personal safety. They have lost confidence in traditional methods and procedures, and in established institutions and professions.

The changes in patterns of employment, the development of careers, the expectations of personal relationships, and the formation of families, together with the uncertainty and insecurity which result from them, are now recognised, well documented and to a large extent understood, if not always accepted. Changes in public institutions – in the structure of government and public service employment, and in the mechanisms of accountability – are also familiar. But the underlying changes in public values, attitudes and expectations, and their consequences for social stability, civic values, and the rights and duties of citizenship, are only now beginning to emerge. At least since the Second World War, the state has been seen as protecting individual freedoms and as providing essential services for its citizens, as a source of authority and as a focus of loyalty. This is the spirit of the International Covenant on Civil and Political Rights, the European Convention on Human Rights and other international instruments. More recently, the state has come to be portrayed not as protecting freedoms but as presenting a threat to them – a threat which may take the form of an expensive and inefficient bureaucracy which consumes a disproportionate share of the nation's wealth; interference in matters of individual choice through excessive regulation or taxation; or even a corrupting influence which destroys self reliance and individual initiative, or even moral integrity, by creating dependence on state benefits. Services provided by the state, as distinct from the private sector, are regarded as inherently wasteful and inefficient.

By the same token, citizens are seen not as people who share a sense of common interest or identity, or who are joined by a sense of mutual obligation and trust, but as individual consumers of public services. Their "rights" are not those of human dignity and respect, but to exercise freedom of choice. The notion of rights has become devalued as selfish or corporatist, and the expectations of citizenship – personal security, work, health, education and housing – have become commodities to be purchased and traded on the most competitive available terms. Public service is judged by the achievement of internal targets and performance indicators, or as the accurate performance of an internal process (perhaps accredited to BS5750), rather than by reference to any social or economic outcomes. The values of public service become those of managerial efficiency, to be achieved for its own sake and not in pursuit of any wider purpose or obligation.

The deconstruction of the state as a source of authority or focus of loyalty has enabled the government, and in effect on the political administration of the day, to take its place. Issues which could once be approached on a basis of evidence or principle are now decided as matters of political conviction or judgement. It is to the political administration that civil servants owe their loyalty. It is the political administration which controls parliament, and which is, therefore, seen as giving the courts and the rest of the apparatus of the state their powers. It is to the political administration, rather than to local communities or related organisations, that public services are held to be primarily accountable. All are required to work in accordance with the centrally and politically determined policies of the government, communicated to them through an ever more elaborate structure of legislation, instructions, objectives, targets, performance indicators, league tables, contracts and service level agreements. The government itself becomes increasingly powerful, centralised and remote, while at the same time it dismantles the state's responsibilities towards its citizens and places such responsibilities as remain in the hands of private or intermediate organisations with their own, often commercial or quasi-commercial, interests and objectives.

These changes are usually presented as common sense, as necessary and even inevitable in a world of global markets, competition and resource constraints, and as uncontroversial and politically neutral. In fact, they may have a profound effect not only on the character of public services but also on the nature of modern society, the relationships within it, and the nature and quality of justice.

Exclusion and inclusion

A parallel analysis distinguishes between what might be termed an "exclusive" and an "inclusive" view of society and citizenship. The "exclusive" view emphasises personal freedom and individual responsibility, but is inclined to disregard the influence of situations and circumstances. It distinguishes between a deserving majority who are self-reliant and law-abiding and entitled to benefit themselves and those around them without interference from others; and an undeserving, feckless, welfare-dependent and even dangerous minority or underclass from whom they need to be protected. The benefits of citizenship, or membership of a community, are confined to those who conform to accepted standards or who can afford to pay for them. They can be withdrawn or withheld at the discretion of officials. Human behaviour is thought to be motivated mainly by a desire for material gain or by fear of punishment or disgrace. There is not much interest in, or respect for, notions of equity or social justice, or of public duty or service. A society which adopts this view is likely to be unsure of itself, suspicious of strangers, hostile towards foreigners and fearful of those who do not conform to its assumptions and stereotypes. It will favour such measures as the carrying of identity cards, the maintenance of personal records – including any criminal convictions – with access to them by those in authority, the reporting of suspicious persons or events, and the large scale use of imprisonment. Its decision-making processes will be cautious, closed and secretive. It will not have much time for open discussion, impartial analysis or rigorous questioning. It will not be much interested in evidence or principle.

The contrasting "inclusive" view is less commonly expressed. It recognises the capacity and will of individuals to change – to improve if they are given guidance, help and encouragement; to be damaged if they are abused or humiliated. It emphasises respect for human dignity and personal identity, and a sense of public duty and social responsibility. It looks more towards putting things right for the future than to allocating blame and awarding punishment, although the latter may sometimes be part of the former. Citizenship and membership of the community are seen as permanent attributes, not to be forfeited by misfortune or failure. The duty to conform to society's or the community's standards is matched by the community's own obligation to support its vulnerable and disadvantaged members. Solutions to social problems or to crime have to be sought by inclusion within the community itself – among parents, in schools, by providing opportunities and hope for young people – and not by punishment or exclusion from it. Authority has to be accountable and it has to be legitimate in the sense that respect for it has to be earned and justified. The "inclusive" view is likely to be characteristic of a society which is open and compassionate, which accommodates and respects diversity, and which has some confidence in the future.

Management and trust

The "exclusive" and "inclusive" views of society have their counterpart in the contrasting "low trust" and "high trust" approaches to the management of organisations. The "low trust" view emphasises the discipline of competition and market forces; the threat of redundancy or dismissal; individual, often short-term, contracts; a top-down structure of output measures and performance indicators; compliance with rules; and material rewards, especially through performance-related pay. Not much is made of such notions as personal loyalty or mutual confidence. This view is characteristic of much of the "new public management". It is associated with the contractual and market based approach to management described earlier in this paper.

The "high trust" view emphasises a participative style of management, based on negotiation, consultation and communication. It rewards teams rather than individuals and it seeks to improve and recognise good performance rather than punish failure. It values equity, discretion and trust. Partnerships and contracts are between equals, not between superiors and inferiors or the more and the less powerful. An organisation's internal style must match the function it has to perform: if public servants, especially those in positions of authority, do not feel respected by their employing organisations or others to whom they feel accountable, they will not easily show respect for others, or receive it in return.

Change and complexity

Throughout the criminal justice system, and in the country as a whole, there is a preoccupation with managing or promoting change. Change may result from social conditions, assumptions or circumstances; it may be initiated by managers or practitioners to improve their services (or just in order to survive); or it may be promoted by government. Preoccupation with change takes different forms and is focused in different ways. Practitioners may be concerned to implement the changes imposed upon them by management or by government, or just to reduce the pressure under which they have to work. Managers and government may be concerned to reduce costs, or to find the means of gaining public support or answering public criticism. All will try to improve the quality of services, but many practitioners and managers feel that their attempts are frustrated or rendered ineffective by the financial, organisational and political pressures to which they have first to respond. Some of them, and many observers and interest groups, may feel that arguments are distorted, facts are misrepresented, priorities are confused and values are corrupted. They may be intimidated by the belief that to express any criticism or to take any risk will be visited by damage to their own careers, loss of funding to their organisations, or attempts at public humiliation.

Practitioners, managers and government have to cope not only with change but also with complexity. Few situations are as straightforward as they may seem at first sight, or as they may be portrayed by the media. Most of those which involve the criminal justice system, whether they are about the treatment of individuals or the development and implementation of policies, involve several organisations and several types of skills. Even within the same professions and occupations, different skills, training and qualifications have been developed for different tasks. The need for different agencies to work together has been acknowledged for at least 15 or 20 years, but practical co-operation, and the shared understanding and sense of purpose to sustain it, are still hard to achieve. Recent examples are to be found in areas as diverse as the treatment of victims, mentally disordered offenders, the mental health of children and adolescents, and the prevention of criminality (Home Office, 1996b; Huskins,

1996; James, 1996; Kurtz, 1996; Audit Commission, 1996).

All these studies show that the management both of change and of complexity requires both reliable evidence and a shared understanding of purposes and principles.

Immediate prospects

Social policy and public service in Britain are now at a critical stage. In many respects, the forces of exclusion and the harsher forms of managerialism seem to be in the ascendant. The political programme for the last session of the present parliament is dominated by proposals for penalisation, punishment and social control. There is increasing demonisation of children, social misfits and those who do not comply with what are portrayed as generally accepted social norms – single parents, the unemployed, the mentally disturbed, refugees, those with a criminal record. In public services, the skills which are rewarded and the values which are respected are increasingly those of accountancy and financial management and of cutting costs. Financial procedures and quantified, internal performance measures take precedence over those which might be concerned with quality or social outcomes. Preoccupation with the short-term extends both forwards, with little interest in any longer term sense of direction or purpose; and backwards, with little or no corporate memory or sense that previous experience might be of value. "Delayering" has removed many of those who might have contributed to a corporate memory, and their loss has weakened the capacity of government or public services to think strategically or to provide support for staff who may feel vulnerable or isolated. Public service is seen not as a career or a long term commitment, but as an episode in an individual's personal advancement to be continued only for as long as it is the most financially rewarding option that is available.

On the other hand, there are some signs of a revival of interest in ideas of citizenship, civic responsibility and social inclusion. There is not yet much agreement on the policies or practical measures which might follow such a revival; and there has been no serious attempt to challenge the more dangerous form of managerialism. But policies concentrating on exclusion and punishment without regard to their operational, financial and social consequences are inherently precarious (for example, in their effect on prisons), and they may come to be seen as disproportionately expensive when set alongside other demands for public expenditure. Appeals to social inclusion and public duty, even to compassion, can still prompt a positive response; calls for prevention rather than punishment can increasingly be heard; and public servants still look for a sense of professional purpose which is based on something more than accountancy and cost effectiveness. The Audit Commission's recent report "Misspent Youth: Young People and Crime" is a serious attempt to base policy on evidence and principle (Audit Commission, 1996).

Policies based on evidence and principle

An evidence-based and principled approach to crime and criminal justice would take existing research, analysis and practical experience as its starting point. Based on that foundation, it would establish a set of principles or parameters which might be on these lines.

Measures to reduce crime and deal with its effects cannot be left to the criminal justice process on its own. Most of the influences on crime and criminality lie elsewhere in society and must be tackled where they are to be found – especially in the experiences of childhood and the opportunities in adolescence.

The purpose of the criminal justice services should not be seen narrowly as controlling and punishing crime, but as public service in a broader sense – protecting the weak and vulnerable, promoting social stability and public safety, upholding standards of justice and civilised behaviour.

This purpose is not always best served by simply enforcing the criminal law, still less by enlarging its scope or increasing its severity. Other methods such as mediation or conciliation may sometimes be more successful. Many situations are not amenable to criminal justice solutions, and it should not be seen as a failure of the system if such a solution is not always practicable.

Victims are entitled to dignity, respect and support. They have special claims on the state and its agencies, and on their fellow citizens, for the experience they have suffered – always for information, support and sympathetic understanding; sometimes for compensation, treatment or counselling; and where necessary for protection from a repeat offence. Their rights or expectations do not however extend to personal influence over the state's decision-making process in respect of an offender, any more than they extend to a general entitlement to the use of force.

Criminals are not a separate class to be controlled or excluded. They should be treated as fellow citizens to be brought so far as possible into a position where they too can undertake the obligations of citizenship and gain the respect of others. Many people will be both offenders and victims at different times in their lives.

Criminal justice services must be efficient and accountable. Much has been achieved in recent years, but efficiency cannot be judged only by means of internal performance indicators without reference to external outcomes, and accountability should not be seen only in terms of financial accountability to central government. A wider vision of accountability is needed, founded on recognised principles and values and secured ultimately by legislation and the rule of law.

A country's approach to crime and criminal justice is a reflection of the attitudes and relationships within society as a whole, and is part of its approach to wider questions of social order and justice and the nature of its institutions and organisations. The approach of a healthy and confident society will be based on values of inclusion and trust; the approach of a society which is divided and unsure of itself will be one of exclusion, suspicion and fear.

Political strength will show itself not in an obsession with punishment but in the courage to seek and analyse information, to listen to argument, and to act rationally on a basis of principle and considered judgement.

The next stage would be to establish a set of aims for the next Parliament, and a programme for achieving them. The aims might be:

- Comprehensive programmes to prevent and reduce crime at local level, effectively coordinated and with stable and flexible funding.
- Support and guidance for children growing up in difficult circumstances and for their parents.
- A juvenile justice system which emphasises prevention, recognises the special nature of childhood and the obligations associated with it, minimises the use of adversarial proceedings and develops alternative procedures where they are appropriate.
- More constructive opportunities for young people entering upon adult life, and help to take advantage of them.
- Support, understanding and protection of victims of crime, with recognition of their corresponding obligations and of the proper limits to their rights and expectations.
- Programmes for offenders whose aim in giving effect to the sentence of the court is to enable them to take their place in society as responsible citizens, and to carry out their obligations and sustain their relationships.
- Structures and procedures designed to promote accountability, a sense of strategic direction, trust and sound working relationships.
- A process of open discussion and consultation, based on rigorous but imaginative research and analysis.

How to influence decision-makers

"How to influence decision-makers" is a question that has troubled academics, practitioners and interest groups for many years. The methods and skills have changed over time, from the civilised but exclusive networks of the English establishment which operated until, roughly, the 1960s to the highly professional but hardly less exclusive, usually expensive and sometimes unscrupulous, lobbying techniques of the 1990s. Many organisations lobby successfully on behalf of their client groups, and several "think tanks" promote new ideas and new approaches, often based on empirical research. But there is a strong sense that voices (at any rate liberal voices) raised on social issues are heard less readily than those representing commercial interests, and the government has been inclined to dismiss representative organisations as self-seeking and not genuinely representative of those on whose behalf they claim to act.

Academics have felt increasingly isolated as the time available for research has become constrained, and funding not only harder to obtain but also subject to requirements which restrict the work to a narrow, short-term, managerial or political agenda. They have been accused of having 'failed' the country for not having found a solution to the problem of crime, or for having contributed so little to the kind of programme which now finds political favour; and of being out of touch with public opinion. They are for their own part inclined to be critical of colleagues whose work they consider to have a political slant, or to be more in the nature of journalism than of scholarship. Obvious means of influencing decision-makers include lobbying Ministers or Members of Parliament, using the media, or educating public opinion. All these methods are available to be used, although there is much frustration over what is seen as lack of interest and support.

One obvious approach is a demonstration of successful schemes and programmes which "work" in the sense of changing behaviour and reducing crime. There is no shortage of possible models, and the problem is more often to sustain the programme, to transfer it to other settings, and to show convincing evidence of its success than to design the model in the first place. And although the emphasis is inevitably on models of specific, practical action, they still need to be set in a coherent intellectual and philosophical framework based on evidence and shared principles. Another approach is to appeal to the financial benefits of preventive and restorative measures as compared with adversarial criminal justice measures leading to conviction and (especially custodial) punishment. A third is to draw attention to the economic cost of measures, or a failure to take measures, which leave large numbers of young people disabled from becoming members of the productive work force. All these are approaches which should appeal to those who take a "market", but not necessarily "exclusive", view of society and human relationships.

Another approach with a more limited, but for many people more powerful, appeal is to argue for a wider sense of social responsibility and citizenship, and to expose features – and long term dangers – of a society which is divided against itself and which devotes more and more of its energy and resources to sustaining that division (Baker and Burnside, 1994; Rutherford, 1996).

It is one thing to think of arguments: it is another to deploy them effectively. Those with access to substantial funding can establish foundations and "think tanks", promote conferences, use professional lobbying techniques, and even form or influence religious groups. All this, and more, has been done by the political right in the United States. For those with fewer resources, the Penal Affairs Consortium is one successful model, and other organisations and individuals could combine in similar ways to develop and promote programmes and ideas and to intervene rapidly and convincingly in public debate as occasion demands. Books can be written and films can be made for general audiences, following the example of Roger Graef (Graef, 1992, 1996).

Once a government was committed to a style of administration which had regard for evidence and principle, it would develop its own machinery and sources of advice and expertise for that purpose. Proposals are often put forward for the appointment of ministers with special responsibilities, or for the creation of new departments, but these will not succeed unless there is the political will to sustain them, and unless they are able to operate successfully in the power structure of Whitehall. The Howard League's neglected report on the Dynamics of Justice makes proposals for a structure of advice and consultation, secured by statute, which may provide a more promising approach (Howard League, 1993).

In the meantime, practitioners, managers, voluntary organisations and academics can do much more to work in partnership not only with each other but with their own communities. They need the confidence, the will and the skills to share information, to listen and explain, to gain trust and to give it in return. They need to think not only in terms of purchasers, providers and customers, but of fellow citizens engaged on a common enterprise and sharing a common purpose. By doing so, they can hope to build more confident, open and inclusive communities and also the public support and interest which will influence government and the media at national levels.

A complete reorientation towards a more active and inclusive sense of citizenship would be a national programme for the twenty first century, but it can begin locally in the twentieth.

Acknowledgement
This paper is developed from, and in some places reproduces, the author's earlier paper "Darkness and Light: Justice, Crime and Management for Today", published by the Howard League for Penal Reform in June 1996.

References

Audit Commission (1996). *Misspent Youth: Young People and Crime.* Audit Commission, London.

Burnside, J. and Baker, N. (1994) *Relational Justice – Repairing the Breach.* Waterside Press, Winchester.

Freedland, M. (1994) *Government by Contract and Public Law.* Spring 1994.

Garland, D. (1996). The Limits of the Sovereign State. Strategies of Crime Control in Contemporary Society. *British Journal of Criminology.* Autumn 1996.

Graef, R. (1992). *Living Dangerously: Young Offenders in their Own Words.* Harper Collins, London.

Graef, R. (1996). *Breaking the Cycle.* Carlton Television, 26 November 1996.

Home Office (1990). Crime, Justice and Protecting the Public. Cm 965 HMSO, London.

Home Office (1996a). *Protecting the Public. The Government's Strategy on Crime in England and Wales.* Cm.3190 HMSO, London.

Home Office (1996b). *The Victim's Charter. A Statement of Service Standards for Victims of Crime.* Home Office, London.

Hood, R. and Shute, S. (1996). *Protecting the Public: Automatic Life Sentences, Parole and High Risk Offender.* Criminal Law Review, November 1996.

Howard League for Penal Reform (1993). *The Dynamics of Justice.* Howard League, London.

Huskins, J. (1996). *Quality Work with Young People: Developing Social Skills and Diversion from Risk.* Youth Clubs UK, London.

James, A. (1996). *Life on the Edge. Diversion and the Mentally Disordered Offender.* Mental Health Foundation, London.

Kurtz, Z. (1996). *Treating Children Well. Guide to Using iithe Evidence Base in Commissioning and Managing Services for the Mental Health of Children and Young People.* Mental Health Foundation, London.

Penal Affairs Consortium (1996). *Protecting the Public: Comments on the White Paper.* Penal Affairs Consortium, London.

Police Foundation/Policy Studies Institute. *The Role and Responsibilities of the Police.* Police Foundation/Policy Studies Institute, London.

Pusey, M. (1991). Economic Rationalism in Canberra. Cambridge University Press.

Rutherford, A. (1996). *Criminal Policy and the Eliminative Ideal.* Inaugural Lecture at Southampton University, 8 October 1996.

Windlesham, Lord (1993). *Responses to Crime*, Vol. 2, Penal Policy in the Making. (1996) Vol. 3, Legislating with the Tide. Clarendon Press, Oxford.

CRIMINAL JUSTICE AND PENAL POLICY IN SCOTLAND
PETER YOUNG

David Faulkner provides a vivid account of how politics affects the formulation and implementation of penal policy. He describes two episodes in the recent history of penal policy in England and Wales; first, how, in the 1980s, a series of policies which were based upon an empirical approach to the understanding of crime, criminal justice and penal policy emerged and, second, how these came to an end in the early 1990's due to important shifts in political culture that took place largely within the Conservative Party. One contrast that Faulkner draws is between the broad thrust of the earlier policies and those which followed them. While the earlier policies aimed to reduce and prevent crime by practical measures which tackled both the circumstances of the offender and the offence, more recent measures are described as explicitly punitive in nature. Faulkner's point is that there is no evidence to support this policy; rather, it appeals to emotions of fear and anxiety; it deals with the criminal by excluding them from society and casting them as the enemy. Faulkner, understandably, portrays this as something of a sad history and he reflects on the considerable organisational and cultural changes that need to occur for a more humane, rational and efficient approach to the control of crime to emerge.

The aim of this note is not to disagree with Faulkner's paper but to add to it by pointing out that much of what he argues either does not apply directly to Scotland, or needs to be modified. This is because Scotland has a separate, distinctive criminal law and this is true also of its penal and criminal justice systems which also differ in many respects from those in England and Wales. Moreover, the organisational, cultural and political context in which policy is made in Scotland is not the same as in England and Wales with the result that, while both systems undoubtedly pursue the same broad objectives of prosecuting and controlling crime, punishing and rehabilitating offenders, they often go about this in distinctive ways, with there being significant and interesting variations in policy in many areas. My aim here is to provide a sketch of some of these key differences from a Scottish point of view.

There is no simple way of characterising the differences between the penal and criminal justice systems of Scotland and England and Wales. In some instances these are fundamental but in others they are more a matter of detail or nuance. An example of a fundamental difference between the two systems can be found in the areas of juvenile justice and in the delivery of social work services to offenders, including probation, where there are basic differences in the design of the institutions and in the policies followed. In both these areas Scotland has followed an explicitly welfare oriented policy. In the case of juveniles this policy is delivered through the unique Childrens' Hearing system. The welfare orientation of Hearing system has been explicit since its inception in the Social Work (Scotland) Act, 1968. Following the recommendations of the Kilbrandon Committee (1964), the Hearings were established as a welfare tribunal, not a criminal court, with the objective of assessing the needs of children in trouble who may require compulsory measures of care. Strictly speaking the Childrens' Hearing is not part of the criminal justice system at all but a part of the local authority social work services department and this institutional separation has been crucial to the continuation of the underlying welfare philosophy. The welfare emphasis of the Hearing system has recently been reaffirmed in the 1995 Children (Scotland) Act. Although this Act introduced important changes in some of the procedures of the Hearing system and has established a new relationship between it and the criminal courts, the Act has kept the core of the system in place. The procedures of the Hearing themselves thus continue to be non-adversarial, its philosophy is one of prevention and cure, not punishment and, by definition, it is premised on the recognition of the special nature of childhood and thus seems close to the ideal juvenile justice system advocated by Faulkner in his final section.

The delivery of social work services for offenders is another area where the Scottish system differs significantly from those in England and Wales. Like the Childrens' Hearing system, the agency responsible for these services is the social work services department of the local government authority. The probation service in Scotland was abolished by the 1968 Social Work (Scotland) Act. As with the Hearing system, the philosophical justification for this emphasised the need to conceive of offending within a framework of welfare and this has continued to the present. There have been shifts and changes, marked by a decline in the belief in traditional concepts of generic social work practice as it applies to offending behaviour but the outlook that replaced this has continued to emphasise on welfare criteria. Importantly, there has been a signal lack of emphasis on the idea of punishment in the community which was a predominant feature of the system in England and Wales in the 1980s and early 1990s. Rather, the proffered outlook in Scotland was one which stressed the effectiveness of certain types of social work intervention in encouraging offenders to address their offending behaviour.

It was the Government that encouraged the adoption of this approach in Scotland in 1989 in a speech delivered by the then Secretary of State for Scotland, Malcolm Rifkind. It is the Government again that has reasserted its support for this philosophy in its recent White Papers, Firm and Fair (1994) and the otherwise more explicitly punitive Crime and Punishment (1996) and in the very important Criminal Justice (Scotland) Act, 1995.

While these examples refer only to two aspects of the Scottish system they have a more general importance and significance. They are evidence both of a difference in the development of criminal justice policy in Scotland which depends, in part, on a difference in institutional structures and also of the extent to which the Scottish system has resisted a drift towards the punitive philosophy that Faulkner describes as now central to the system in England and Wales. They are not the only example which could be given of this. Another can be found in Scottish prisons. The basic statement of the present policy followed by the Scottish Prison Service (SPS) is contained in Opportunity and Responsibility (1990), a paper published by SPS, which describes what has come to be known as the 'opportunity' agenda. The underlying idea of the opportunity agenda begins from the premise that imprisonment is about punishing individuals but then argues that the role of the prison should be to offer the responsible prisoner a series of opportunities, through counselling and other forms of training, to address the causes of his or her offending behaviour. The consequence of the implementation of this policy is that prisons in Scotland now provide a greater range of what have come to be called 'positive' regimes than they did when rehabilitation was the prevailing official philosophy. In some ways this policy does amount to a belief that the 'prison works', but not in the punitive sense that this has come to be understood in England and Wales and the USA.

While other examples of the resistance of the Scottish system toward a punitive philosophy could be given, it would be wrong to create the impression that this philosophy is absent altogether from the Scottish scene. The appointment of Mr Forsyth as Secretary of State has shifted policy in that direction and the current Crime Bill (Scotland) is a testament to this. If the Bill becomes an Act and is implemented, it will bring the Scottish and English systems closer together in some areas of policy, especially in sentencing. A careful reading of the White Paper informing the Bill, Crime and Punishment, would show that there continues to be a commitment to certain aspects of the welfare philosophy described above. In addition the institutional structures both within the criminal justice system and in the government that support this philosophy will remain in place. Effectively, for a large part of the twentieth century, the making of penal and criminal justice policy has been in the hands of a policy-making network that works through Edinburgh, not London. The position of this policy-making network has been an important aspect of the delicate constitutional balance that has characterised the relationship between Scotland and England and Wales. The outcome of the changes most likely to affect this balance in the near future will probably lead not to an ironing out of the differences between Scotland and England and Wales but to their growth.

ALIENS FROM SPACE?

PATRICK MOORE

Patrick Moore CBE FRAS was born in 1923 and has been inspiring young people with his enthusiasm for astronomy for more than 40 years. As presenter of "The Sky at Night" since 1957 he is a familar face to all amateur astronomers and is a well known broadcaster on all matters astronomical. He has written more than 60 books and has edited The Yearbook of Astronomy *since 1962. His many achievements have been recognised by Astronomical Societies from all around the world.*

Is there life on other worlds? This is a question which has been asked time and time again – and to which we have, as yet, no final answer. However, we have learned a great deal in recent years, and even if we still lack proof we are at least in a position to make intelligent guesses.

First, we must decide just what is meant by "life". We have to admit that we are still not quite sure, but we know a great deal about living matter, and so far as we can tell the key to the whole situation is carbon. Only atoms of carbon can band together with other atoms to form the large, complicated molecules needed for life. Silicon can make a gallant attempt, but there is no evidence of silicon-based life anywhere, and we can also be confident that even the remotest galaxies are made up of the same elements as those we find on Earth. In short, there is every reason to assume that life, wherever it may be found, must be carbon-based.

Of course, there could be a fundamental flaw in this argument, opening the door for totally alien forms of the type known to science-fiction writers as BEMs (Bug-Eyed Monsters). Yet if this is true, then most of our modern science is wrong, and of this there is not a shred of evidence. Faced with a set of facts which may not be complete, all we can do is to interpret them in the most reasonable way possible. To give an analogy; it is generally believed that the Sun will rise in the east tomorrow morning. It could, of course, rise in the north, but this would destroy all modern science. Therefore, until the Sun really does rise in the north, we must keep to accepted science; and for the same sort of reason, we can discount Bug-Eyed Monsters.

This is not to say that life-forms elsewhere need look like us; there is not much outward resemblance between a man and an earwig – but both are carbon-based. Appearances can be deceptive!

For the moment, then, let us confine ourselves to "life as we know it", and see whether we have any hope of tracking any living thing beyond the Earth. Clearly we must start with the Solar System, which is the only part of the universe which we can see in real detail and which we can explore with our spacecraft – so what are the prospects?

It is strange now to reflect that Sir William Herschel, discoverer of the planet Uranus and arguably the greatest of all astronomical observers, was convinced that even the Sun was inhabited; he never changed his mind, and his faith in Sun dwellers was unshaken at the time of his death in 1822. Rather less implausible was the idea of life on the Moon; in the mid-nineteenth century a famous German observer, Franz von Paula Gruithuisen, announced the discovery of a lunar city with "dark gigantic ramparts", and in the 1830s many people were taken in by a series of hoax articles in a New York paper reporting observations of various weird and wonderful lunar creatures. But the Moon has virtually no atmosphere, and we can certainly rule out airless worlds as suitable for life. Of course, no chances can be taken; the first Apollo astronauts were quarantined on their return, but it soon became clear that the Moon is, and always has been, absolutely sterile.

Of the planets, Mercury is ruled out because of its lack of atmosphere, but Venus and Mars cannot be dismissed so easily, and before the Space Age it was widely believed that life might flourish there. Venus hid its secrets well. Though it is slightly smaller and less massive than the Earth, it has a dense, cloud-laden atmosphere which hides its surface completely; there is no such thing as a sunny day on Venus! In the 1920s, Svante Arrhenius, the Swedish scientist whose work was good enough to win him a Nobel Prize, was claiming that Venus must be in a state similar to that of the Earth in Carboniferous times, with ferns, horse-tails, marshes and creatures such as dragonflies. If this had been the case, then there seemed no reason why life on Venus should not evolve in the same way as it has done on Earth. Later, American astronomers suggested that most of the surface was covered with water. This would have led to a rather bizarre situation. It was already known that the atmosphere of Venus is rich in carbon dioxide; if there had been oceans, the water would have been fouled by the atmospheric CO_2 – producing seas of soda-water (though it did not seem that there was much chance of finding any whisky to mix with it).

Alas! we now know better. The atmosphere is almost pure carbon dioxide; the surface temperature approaches 1000 degrees Fahrenheit; the atmospheric pressure is about 90 times that of our air at sea-level, and those beautiful clouds are rich in sulphuric acid. Go to Venus, step out of your spacecraft, and you will be promptly suffocated, fried, squashed and corroded. Though Venus is named after the Goddess of Beauty, conditions there are remarkably like the conventional picture of hell.

Why are Venus and Earth so different? The reason must be that Venus is closer to the Sun – on average 67,000,000 miles out, as against our 93,000,000 miles. It may be that in the early days of the Solar System, when the Sun was less luminous that it is now, Venus and the Earth started to evolve along the same lines; but as the Sun became hotter, the oceans of Venus boiled away, the carbonates were driven out of the rocks, and in a short time – cosmically speaking – Venus changed from a potentially life-bearing world into the furnace-like inferno of today. It is sobering to reflect that if the Earth had been a mere 25,000,000 miles closer to the Sun, life could never have survived, and there would have been no Edinburgh Festival.

Mars, then? Here the situation is different; Mars is smaller than the Earth, and further from the Sun (141,500,000 miles on average), with a thin atmosphere which is (usually) transparent enough to let us examine the true surface. Percival Lowell, who set up a major observatory in Arizona in 1896, and equipped it with a large telescope, was convinced that he had observed artificial canals on the planet, presumably built by the Martians to form an irrigation network on a world which was desperately short of water; in his last book Lowell wrote "That Mars is inhabited by beings of some sort or other is as certain as it is uncertain what these beings may be". And it is true that if Lowell's canals had existed, Mars would have been inhabited. Unfortunately, we now know that they were tricks of the eye. Space pictures have shown that Mars is a world of craters, valleys, plains and massive volcanoes which may or may not be finally extinct. Yet there are also features which look so like dry riverbeds that it is difficult to interpret them in any other way, in which case running water must have existed there before the atmosphere became too thin. (We cannot finally prove that the features are riverbeds, but remember what was said at a recent conference: "If it looks like a duck, quacks like a duck and waddles like a duck, then maybe it *is* a duck.")

If Mars were once much less hostile than it is now, then life may well have appeared there, to die out or 'go underground' when the conditions deteriorated. In 1996, there was a suggestion that certain meteorites, found in Antarctica, might be of Martian origin, and contained evidence of past life. This is an interesting possibility, but no more. We have no definite proof that the meteorites came from Mars, and neither is there definite proof that the observed features really do indicate former life. Probably we will not know until an unmanned sample-and-return probe manages to bring us samples which are unquestionably Martian. This should be possible within the next few years.

If we do find traces of former life on Mars, the implications will be of tremendous importance; we will know that life will appear wherever conditions are suitable for it, and will develop as far as conditions allow. In the case of Mars, there can never have been anything as advanced as a dandelion, and certainly no canal-building Martians, but it is certainly true that it is less unlike the Earth than any other planet in the Sun's family, and it will indeed be strange if permanent colonies are not set up there during the twenty-first century.

This leads on to another interesting speculation. Journeys to the Moon are brief affairs, but it takes much longer to reach Mars, and the first expeditions will not be quick 'there and back' missions of the Apollo type. When Bases are established, there will be men, women, and children. But on Mars the force of gravity is only one-third that on Earth. Will a boy or girl born and brought up on Mars ever be able to come to Earth? Imagine how you would feel if, without any extra strength in your muscles, you suddenly increased your weight by a factor of three. You might not be able to cope – and it is conceivable than in the foreseeable future there will be two distinct branches of *Homo sapiens*: Earthmen, and our Martian relatives who are barred from visiting the planet of their ancestors, though naturally they would be quite happy on the Moon. This sounds like science fiction in 1996, but it may have become science fact by 2096.

The other members of the Sun's family are less promising. The smaller bodies have no atmospheric coating, and the giants have gaseous surfaces as well as being bitterly cold (and in most cases surrounded by zones of lethal radiation). Europa, one of Jupiter's satellites, has a smooth, icy surface, and it has been suggested that there may be a subterranean ocean of ordinary water warm enough to support life, but I admit that I take this idea with a very large grain of cosmic salt. Titan, the senior attendant of Saturn, has a dense atmosphere rich in nitrogen, but there is also a great deal of methane, and the surface temperature is intolerably low. The ingredients for life exist in Titan, but the actual chances of finding life there seem to be minimal. We should know more in 2004, with the landing of a special probe named in honour of Huygens, who discovered Titan in 1655. Whether the probe will land on solid ground, or plunge into a chemical ocean, remains to be seen.

En passant, there will come a time – several thousands of millions of years hence – when the Sun will become more powerful, and the Earth must suffer a sad fate. Titan will then be warmed. Unfortunately this means that its atmospheric molecules will speed up and break free, so that there is no chance of humanity emigrating there! Moreover, the Sun's highly luminous stage will not persist, and will be succeeded by a period when the Sun is reduced to a small, dense, feeble white dwarf star, still circled by the ghosts of its surviving planets.

There have been many suggestions that life in Earth did not originate here, but was brought to our world via a comet or comets. Again, proof is lacking one way or the other, and we have to admit that our knowledge of the actual origin of life is very meagre.

All in all, the chances of life in the Solar System beyond Earth do not seem very high, but we must not be parochial. Our Galaxy contains about a hundred thousand million stars, of which the Sun is only one; moreover the Sun is a very ordinary kind of star – and our Galaxy is a very ordinary kind of galaxy; there are thousands of millions of others. In all this host it is surely absurd to believe that only our Sun is attended by a peopled planet. What hope have we of contacting other intelligent beings, always assuming that they really exist?

First, we must try to establish that planetary systems are common. A few decades ago it was widely believed that the planets in the Solar System were pulled off the Sun by the action of a passing star, and if this had been true then similar systems would have been rare; the stars are widely separated in space, and even "close encounters" can seldom occur. Now, however, we think differently. We believe that planets were formed from material spread round the youthful Sun, and what can happen to the Sun can happen to other stars also. Indeed, pictures taken with the Hubble Space Telescope show that there are many stars with significant disks around them. There is other evidence, too. In 1983 the Infra-Red Astronomical Satellite, IRAS, found that some stars – such as the brilliant Vega, and the southern Beta Pictoris – are associated with cool material which could well be planet-forming; and in the case of Beta Pictoris, this material has actually been imaged telescopically.

What we cannot do – at least, not yet – is to see other planets directly; they are too small, too faint and too close to their parent stars. However, recent work shows that some stars are "wobbling" slightly, presumably because they are being pulled around by orbiting bodies which are probably planets. Everything is starting to indicate that planetary systems are even more common than was believed a few years ago.

The limitation is that all these methods are valid only for very large planets, more like Jupiter than the Earth in mass. But if there are any planets at all, then Earths may confidently be expected – and if we have a planet similar to the Earth, moving round a star similar to the Sun, then why should there not be life similar to ours? It sounds eminently reasonable.

At the moment, the only practicable way of establishing contact seems to be by radio, since radio waves travel at the same speed as light (186,000 miles per second). If we could pick up signals which were sufficiently rhythmical to be interpreted as non-natural, then the main question would be answered. The two nearest stars which are sufficiently solar in type to be regarded as potential planetary centres are known as Tau Ceti and Epsilon Eridani; both are about 11 light-years away. (One light-year is the distance travelled by light in a year; approximately 5.8 million million miles.) Efforts have been made to 'listen out' at selected wavelengths to see if any artificial signals can be received from either of these systems. The first attempt dates back as far as 1960; since then we have had the SETI programme (Search for Extra-Terrestrial Intelligence), which was initiated in America with great enthusiasm, but was unfortunately cancelled by Congress a year later, on the grounds of expense.

Up to now the results have been negative, but the chances are not nil, and the searches will continue. Obviously we must use the language of mathematics. We did not invent mathematics; we merely discovered it, and presumably other races will have done the same.

There is another limitation. Suppose that we send a message to Tau Ceti in 1996? It will arrive in 2007; if some obliging radio astronomer replies straight away, we will have a reply in 2018 – a total delay of 22 years. Quick-fire repartee will be rather difficult.

Actual travel is out of the question with our present technology. Rockets would take far too long; devices such as space-arks seem to belong to the realm of science fiction, and when we consider time-warps, space-warps, teleportation and thought-travel we have entered the domain of Lord Darth Vader and the star-ship *Enterprise*. Yet science fiction does have a habit of turning into science fact; even air travel would have seemed equally fantastic a few centuries ago, or even less. One day we may master these exotic forms of interstellar travel, and we may be no further away from them in time than King Canute was from television.

There is always a chance that the mountain will come to Mahomet rather than *vice versa*. Alien visitations cannot be ruled out, but there is no evidence that they have happened yet, despite the claims of the flying saucer enthusiasts. Naturally, it has been claimed that there have been visitations, and that there have been massive 'cover-up' operations by world governments – but we have to admit that conspiracy theories are characteristic of the honest but misguided crank!

There is, however, one more point which seems to be well worth discussing. It has been tacitly assumed that alien beings must be malevolent; this stems in part from the classic story *The War of the Worlds* by H.G. Wells, in which the Earth is invaded by grotesque monsters from Mars. Wells has had many imitators, and the pulp magazines of the 1990s are not so very different from those of the 1930s, which were crawling with bug-eyed monsters intent upon destroying Earth and all its inhabitants. There have even been serious and sober scientists who have objected to our sending out radio messages, or allowing space-probes such as the Voyagers to leave the Solar System, on the grounds that it would be dangerous to draw attention to ourselves. This seems to be a very short-sighted view, on two grounds. First, we have been broadcasting now for at least sixty years, so that to any race within sixty light-years of us we are "radio noisy"; we could not hide ourselves even if we wanted to do so. Secondly, why should aliens be hostile? Any race which is capable of interstellar travel will be far more advanced than we are, and will have put the folly of war far behind them. This means that they will come in a spirit of friendship rather than enmity. How we would ourselves react is another matter, but it is significant that at a recent General Assembly of the International Astronomical Union a whole day was devoted to the recommended procedure in the event of contact being made with any being from afar.

Our views have changed in recent times. No longer do we regard ourselves as important; no doubt there are many races which are far superior to ourselves. There may also be ruined, radioactive worlds whose inhabitants have destroyed themselves by warfare, but we must hope that in most cases sanity will prevail. For the moment, all we can really do is to keep searching.

I think I must end by answering a question which was put to me in a radio interview not so long ago. We had been talking about the possibility of contact with an alien race, and the interviewer asked me what I would say if a flying saucer landed in my garden and a little green man from Mars stepped out. I knew exactly what my response would be. I would say: "Good-morning. Please come with me to the nearest television studio!"

NOT INVENTED HERE

IAIN M. BANKS

Born in 1954 in Dunfermline, Fife, Iain Banks went to Stirling University and was variously employed before giving up his day job in 1984 when The Wasp Factory *was published. He writes mainstream books under the above name and puts an "M" (for Menzies) in when he's writing Science Fiction.*

HUP! ... and here we are, waking up. Quick scan around, nothing immediately threatening, it would seem ... Hmm. Floating in space. Odd. Nobody else around. That's funny. View's a bit degraded. Oh-oh, that's a bad sign. Don't feel quite right, either. Stuff missing here ... Clock running way slow, like it's down amongst the electronics crap ... Run full system check.

... Oh, good grief!

The drone drifted through the darkness of interstellar space. It really was alone. Profoundly, even frighteningly alone. It picked through the debris that had been its power, sensory and weapon systems, appalled at the wasteland it was discovering within itself. The drone felt weird. It knew who it was – it was Sisela Ytheleus 1/2, a type D4 military drone of the Explorer Ship *Peace Makes Plenty*, a vessel of the Stargazer Clan, part of the Fifth Fleet of the Zetetic Elench – but its real-time memories only began from the instant it had woken up here, a zillion klicks from anywhere, slap bang in the middle of nothing with the shit kicked out of it. What *a mess!* Who had done this? What had *happened* to it? Where were its memories? Where was its mind-state?

Actually it suspected it knew. It was functioning on the middle level of its five stepped mind-modes; the electronic.

Below lay an atomechanical complex and beneath that a bio-chemical brain. In theory the routes to both lay open; in practice both were compromised. The atomechanical mind wasn't responding correctly to the system-state signals it was receiving, and the biochemical brain was simply a mush; either the drone had been doing some hard manoeuvring recently or it had been clobbered by something. It felt like dumping the whole biochemical unit into space now but it knew the cellular soup its final back-up mindsubstrate had turned into might come in handy for something.

Above, where it *ought* to be right now, there were a couple of enormously wide conduits leading to the photonic nucleus and beyond that the true AI core. Both completely blocked off, and metaphorically plastered with warning signals. The equivalent of a single lit telltale adjacent to the photonic pipe indicated there was activity of some sort in there. The AI core was either dead, empty or just not saying.

The drone ran another systems-control check. It *seemed* to be in charge of the whole outfit, what was left of it. It wondered if the sensor and weaponry systems degradation was real. Perhaps it was an illusion; perhaps those units were in fact in perfect working order and under the control of one or both of the higher mind components. It dug deeper into the units' programming. No, it didn't look possible.

Unless the whole situation was a simulation. That was possible. A test: what would you do if you suddenly found yourself drifting alone in interstellar space, almost every system severely damaged, reduced to a level-three mind-state with no sign of help anywhere and no recollection how you got here or what happened to you? It *sounded* like a particularly nasty simulation problem; a nearly-worst-case scenario dreamt up by a Drone Training and Selection Board.

Well, there was no way of telling, and it had to act as though it was all real.

It kept looking around inside its own mind-state. *Ah ha.*

There were a couple of closed sub-cores intact within its electronic mind, sealed and labelled as potentially – though not probably – dangerous. There was a similar warning attached to the self-repair control-routine matrices. The drone let those be for the moment. It would check out everything else that it could before it started opening packages with what might prove to be nasty surprises inside.

Where the hell *was* it? It scanned the stars. A matrix of figures flashed into its consciousness. Definitely the middle of nowhere. The general volume was called the Upper Leaf-Swirl by most people; forty-five kilolights from galactic centre. The nearest star – fourteen standard light months away – was called Esperi, an old red giant which had long since swallowed up its complement of inner planets and whose insubstantial orb of gases now glowed dully upon a couple of silent, icy worlds and a distant cloud of comet nuclei. No life anywhere; just another boring, barren system like a hundred million others.

The general volume was one of the less well-visited and relatively uninhabited regions of the galaxy. Nearest major civilisation point; the Sagraeth system, forty light years away, with a stage-three lizardoid civilisation first contacted by the Culture a decade ago. Nothing special there. Voluminal influences/interests rated Creheesil 15%, Affront 10%, Culture 5% (the normal claimed minimum, the Culture's influence/interest equivalent of background radiation), and a smattering of investigations and flybys by twenty other civilisations making up a nominal 2%; otherwise not a place anybody was really interested in; a two-thirds forgotten, disregarded region of space. Never before directly investigated by the Elench, though there had been the usual deep-space remote scans from afar, showing nothing special. No clues there.

Date; n4.28.803, by the chronology the Elench still shared with the Culture. The drone's service log abstract recorded that it had been built as part of a matched pair by the *Peace Makes Plenty* in n4.13, shortly after the ship's own construction had been completed.

Most recent entry; '28.725.500: ship leaving Tier habitat for a standard sweep-search of the outer reaches of the Upper Leaf-Swirl. The detailed service log was missing. The last flagged event the drone could find in its library dated from '28.802; a daily current affairs archive update. So had that been just yesterday, or could something have happened to its clock?

It scrutinised its damage reports and searched its memories. The damage profile equated to that caused by plasma fire, and – from the lack of obvious patterning – either an enormous plasma event very far away or plasma fire – possibly fusion-sourced – much closer but buffered in some way. A nearby plasma implosure was the most obvious example. Not something it could do itself. The ship could, though.

Its X-ray laser had been fired recently and its field-shields projectors had soaked up some leak-through damage. Consistent with what would have happened if something just like itself had attacked it. *Hmm.* One of a matched pair.

It thought. It searched. It could find no further mention of its twin.

It looked about itself, gauging its drift, and searching.

It was drifting at about two-eighty klicks a second, almost directly away from the Esperi system. In front of it – it focused all its damaged sensory capacity to peer ahead – nothing; it didn't appear to be aimed *at* anything.

Two-eighty klicks a second; that was somewhere just underneath the theoretical limit beyond which something of its mass would start to produce a relativistic trace on the surface of space-time, if one had perfect instrumentation. Now, was that a coincidence, or not? If not, it might have been slung out of the ship for some reason; Displaced, perhaps. It concentrated its senses backwards. No obvious point of origin, and nothing coming after it, either. Hint of something though.

The drone refocused, cursing its hopelessly degraded senses. Behind it, it found ... gas, plasma, carbon. It widened the cone of its focus.

What it had discovered was an inflating shell of debris, drifting after it at a tenth of its speed. It ran a rewind of the debris shell's expansion; it originated at a point forty klicks behind the position where it had first woken up, eighteen fifty-three milliseconds ago.

Which implied it had been drifting totally unconscious for nearly half a second. *Scary.*

It scanned the distant shell of expanding particles. They'd been hot. Messy. That was wreckage. Battle wreckage, even. The carbon and the ions could originally have been part of itself, or part of the ship, or even part of a human. A few molecules of nitrogen and carbon dioxide. No oxygen.

But all of it doing just 10% of its own velocity. Odd, that. As though it had somehow been prioritised out of a sudden appearance of matter. Again, as though it had been Displaced, perhaps.

The drone flicked part of its attention back inside, to the sealed cores in its mind substrate with their warning notices. Can't put this off any longer, I suppose, it thought.

It interrogated the two cores. *PAST,* the first was labelled. The other one was simply called 2/2.

Uh-huh, it thought.

It opened the first core and found its memories.

II

Genar-Hofoen floated within the shower, buffeted from all sides by the streams of water. The fans sucking the water back out of the AG shower chamber sounded awfully loud this morning. Part of his brain told him he was running short of oxygen; he'd either have to leave the shower or grope for the air hose which was probably in the last place he'd feel for it. It was either that or open his eyes. It all seemed too much bother. He was quite comfortable where he was.

He waited to see what would give first.

It was his brain's indifference to the fact he was suffocating. Suddenly he was wide awake and flailing around like some drowning basic-human, desperate for breath but afraid to breathe in the constellation of water globules he was floating within. His eyes were wide open. He saw the air hose and grabbed it. He breathed in. Shit it was bright. His eyes dimmed the view. That was better.

He felt he'd showered enough. He mumbled, 'Off, off', into the air hose mask a few times, but the water kept on coming. Then he remembered that the module wasn't talking to him right now. because he'd told the suit to accept no more communications last night. Obviously such irresponsibility had to be punished by the module being childish. He sighed.

Luckily the shower had an Off button. The water jets cut off. Gravity was fed gently back into the chamber and he floated slowly down with the settling blobs of water. A reverser field clicked on and he looked at himself in it while the last of the water drained away, sucking in his belly and sticking out his chin while he turned his face to the best angle and smoothed down a few upstart locks of his blond curls.

'Well, I may feel like shit but I still look great', he announced to nobody in particular. For once, probably even the module wasn't listening.

* * *

'Sorry to force the pace', the representation of his uncle Tishlin said.

''s all right', he said through a mouthful of *feyl* steak. He washed it down with some warmed-over infusion the module had always assured him was beneficial when you hadn't had enough sleep. It tasted disgusting enough to be either genuinely good for you, or just one of the module's little jokes.

'Sleep okay?' his uncle's image asked. He was, apparently, sitting across the table from Genar-Hofoen in the module's dining room, a pleasantly airy space filled with porcelain and flowers and boasting a seemingly real-time view on three sides of a sunlit mountain valley, which in reality was half a galaxy away. A small serving drone hovered near the wall behind the man.

'Good two hours', Genar-Hofoen said. He supposed he could have stayed awake the night before when he'd first discovered his uncle's hologram waiting for him; he could have glanded something to keep him bright and awake and receptive and got all this over with then, but he'd known he'd end up paying for it eventually and besides, he wanted to show them that just because they'd gone to the trouble of persuading his favourite uncle to record a semantic-signal-mind-abstract-state or whatever the hell the module had called it, he still wasn't going to jump just because they said so. The only concession he'd made to all the urgency was deliberately not to dream; he had a whole suite of pretty splendid dream-accessible scenarios going at the moment, several of them incorporating some powerfully good and satisfying sex, and it was a positive sacrifice to miss out on any of them.

So he'd gone to bed and had a pretty good if maybe still not quite long enough sleep and Uncle Tishlin's message had just had to sit twiddling its abstract semantics in the module's AI core, waiting till he got up.

So far all they'd done was exchange a few pleasantries and talk a little about old times; partly, of course, so that Genar-Hofoen could satisfy himself that this apparition had genuinely been sent by his uncle and SC had paid him the enormous compliment of sending not one but two personality-states to him in order to argue him round to doing whatever it was they wanted from him (that the hologram might be a brilliantly researched forgery created by SC would be even more of a compliment ... but that way lay paranoia).

'I take it you had a good evening', Tishlin's simulation said.

'Enormous fun'.

Tishlin looked puzzled. Genar-Hofoen watched the expression form on his uncle's face and wondered how comprehensive was the duplication of his uncle's personality now encoded – living, if you wanted to look at it that way – in the module's AI core. Did whatever was in there – sent here enciphered with the specific task of persuading him to cooperate with Special Circumstances – actually *feel*? Or did it just appear to?

Shit, I must be feeling bad, Genar-Hofoen thought. I haven't bothered about that sort of shit since university.

'How can you have enormous fun with... aliens?' the hologram asked, eyebrows gathering.

'Attitude', Genar-Hofoen said cryptically, slicing off more steak.

'But you can't drink with them, eat with them, can't really touch them, or want the same things ... '. Tishlin said, still frowning.

Genar-Hofoen shrugged. 'It's a kind of translation', he said. 'You get used to it'. He munched away for a moment while his uncle's program – or whatever it was – digested this. He pointed his knife at the image. '*That's* something I'd want, in the unlikely event I agree to do whatever it is they want me to do'.

'What?' Tishlin said, leaning back, arms crossed.

'I want to become an Affronter'.

Tishlin's eyebrows elevated. 'You want *what*, boy?' he said.

'Well, some of the time', Genar-Hofoen said, half turning his head to the drone behind him; the machine came quickly forward and refilled his glass with the infusion. 'I mean, all I want is an Affronter body, one that I can just Sort of zap into and ... well, just *be* an Affronter. You know; socialise. I don't see what the problem is, really. In fact I keep telling them it'll be a great thing for Culture-Affront relations. I'd really be able to relate to these guys; I could really be one of them. Hell; isn't that what this ambassador shit is supposed to be all about?' He belched. 'I'm sure it could be done. The module says it could but it shouldn't and says it's asked elsewhere and I know all the standard objections, but I think it'd be a great idea. I'm damn sure *I'd* enjoy it, I mean I could always sort of zap back into my own body anytime ... this is really shocking you, isn't it, Uncle?'

The image shook its head. 'You always were the oddest child, Byr. I suppose I should have known what to expect from you. Anybody who'd go out there to live with the Affront in the first place has to be slightly strange'.

Genar-Hofoen held his arms out wide. 'But I'm just doing what you did!' he protested.

'I only wanted to *meet* weird aliens, Byr; I didn't want to become one of them'.

'Heck, and I thought you'd be proud of me'.

'Proud but worried. Byr, are you seriously suggesting that becoming an Affronter would be part of your price for doing what SC asks?'

'Certainly', Genar-Hofoen said, and squinted up at the hammer-beamed ceiling. 'I vaguely recall asking for a ship as well last night and the *Death And Gravity* saying yes ...' he shook his head and laughed. 'Must have imagined it'. He finished the last of the steak.

'They've told me what they're prepared to offer, Byr', Tishlin said. 'You didn't imagine it'.

Genar-Hofoen looked up. 'Really?' he asked.

'Really', Tishlin said.

Genar-Hofoen nodded slowly. 'And how did they persuade you to act as go-between, Uncle?' he asked.

'They only had to ask, Byr. I may not be in Contact any more but I'm happy to help out when I can, when they have a problem'.

'This isn't Contact, Uncle, this is Special Circumstances', Byr said quietly. 'They tend to play by slightly different rules'.

Tishlin looked serious; the image sounded defensive. 'I know that, boy. I asked around some of my contacts before I agreed to do this; everything checks out, everything seems to be ... reliable. I suggest you do the same, obviously, but from what I can see, what I've been told is the truth'.

Genar-Hofoen was silent for a moment. 'Okay. So what have they told you, Uncle?' he asked, draining the last of the infusion. He frowned, wiped his lips and inspected the napkin. He looked at the sediment in the bottom of the glass, then glared at the servant drone. It wobbled in the drone equivalent of a shrug and took the glass from his hand.

Tishlin's representation sat forward, putting its arms on the table. 'Let me tell you a story, Byr'.

'By all means', Genar-Hofoen said, picking something from his lips and wiping it on the napkin. The serving drone started to remove the rest of the breakfast things.

'Long ago and far away – two and a half thousand years ago', Tishlin said, 'in a wispy tendril of suns outside the Galactic plane, nearest to Asatiel Cluster, but not really near to that or anywhere else – the *Problem Child,* an early General Contact Unit, Troubadour Class, chanced upon the ember of a very old star. The GCU started to investigate. And it found not one but two unusual things'.

Genar-Hofoen drew his gown about him and settled back in his seat, a small smile on his lips. Uncle Tish had always liked telling stories. Some of Genar-Hofoen's earliest memories were of the long, sunlit kitchen of the house at Ois, back on Seddun Orbital; his mother, the other adults of the house and his various cousins would all be milling around, chattering and laughing while he sat on his uncle's knee, being told tales. Some of them were ordinary children's stories – which he'd heard before, often, but which always sounded better when Uncle Tish told them – and some of them his uncle's own stories, from when he'd been in Contact, travelling the galaxy in a succession of ships, exploring strange new worlds and meeting all sorts of odd folk and finding any number of weird and wonderful things amongst the stars.

'Firstly', the hologram image said, 'the dead sun gave every sign of being absurdly ancient. The techniques used to date it indicated it was getting on for a trillion years old'.

'What?' Genar-Hofoen snorted.

Uncle Tishlin spread his hands. 'The ship couldn't believe it either. To come up with this unlikely figure, it used ...' the apparition glanced away to one side, the way Tishlin always had when he was thinking, and Genar-Hofoen found himself smiling, '...isotopic analysis and flux-pitting assay'.

'Technical terms', Genar-Hofoen said, nodding. He and the hologram both smiled.

'Technical terms', the image of Tishlin agreed. 'But no matter what it was they used or how they did their sums, it always came out that the dead star was at least fifty times older than the universe'.

'I never heard that one before', Genar-Hofoen said, shaking his head and looking thoughtful.

'Me neither', Tishlin agreed. 'Though as it turns out it was released publicly, just not until long after it had all happened. One reason there was no big fuss at the time was that the ship was so embarrassed about what it was coming up with it never filed a full report, just kept the results to itself, in its own mind'.

'Did they have proper Minds back then?'

Tishlin's image shrugged. 'Mind with a small "m"; AI core, we'd probably call it these days. But it was certainly sentient and the point is that the information remained in the ship's head, as it were'.

Where, of course, it would remain the ship's. Practically the only form of private property the Culture recognised was thought, and memory. Any publicly filed report or analysis was theoretically available to anybody, but your own thoughts, your own recollections – whether you were a human, a drone or a ship Mind – were regarded as private. It was considered the ultimate in bad manners even to think about trying to read somebody else's – or something else's – mind.

Personally, Genar-Hofoen had always thought it was a reasonable enough rule, although along with a lot of people over the years he'd long suspected that one of the main reasons for its existence was that it suited the purposes of the Culture's Minds in general, and those in Special Circumstances in particular.

Thanks to that taboo, everybody in the Culture could keep secrets to themselves and hatch little schemes and plots to their hearts' content. The trouble was that while in humans this sort of behaviour tended to manifest itself in practical jokes, petty jealousies, silly misunderstandings and instances of tragically unrequited love, with Minds it occasionally meant they forgot to tell everybody else about finding entire stellar civilisations, or took it upon themselves to try to alter the course of a developed culture everybody already did know about (with the almost unspeakable implication that one day they might do just that not with a culture but with *the* Culture ... always assuming they hadn't done so already, of course).

'What about the people on board the Culture ship?' GenarHofoen asked.

'They knew as well, of course, but they kept quiet, too. Apart from anything else, they had *two* weirdnesses on their hands; they assumed they had to be linked in some way but they couldn't work out how, so they decided to wait and see before they told everybody else'. Tishlin shrugged. 'Understandable, I suppose; it was all so outlandish I suppose anybody would think twice about shouting it to the rooftops. You couldn't get away with such reticence these days, but this was then; the guidelines were looser'.

'What was the other unusual thing they found?'

'An artifact', Tishlin said, sitting back in the seat. 'A perfect blackbody sphere fifty klicks across, in orbit around the unfeasibly ancient star. The ship was completely unable to penetrate the artifact with its sensors, or with anything else for that matter, and the thing itself showed no signs of life. Shortly thereafter the *Problem Child* developed an engine fault – something almost unheard of, even back then – and had to leave the star and the artifact. Naturally, it left a load of satellites and sensor platforms behind it to monitor the artifact; all it had arrived with, in fact, plus a load more it had made while it was there.

'However, when a follow-up expedition arrived three years later – remember, this all happened on the galactic outskirts, and speeds were much lower then – it found nothing; no star, no artifact, and none of the sensors and remote packages the *Problem Child* had left behind; the outgoing signals apparently coming from the sentry units stopped just before the follow-up expedition arrived within monitoring range. Ripples in the gravity field near by implied the star and presumably everything else had vanished utterly the moment the *Problem Child* had been safely out of sensor range'.

'Just vanished?'

'Just vanished. Disappeared without trace', Tishlin confirmed. 'Most damnable thing, too; nobody's ever just lost a sun before, even if it was a dead one.

'In the meantime, the General Systems Vehicle which the *Problem Child* had rendezvoused with for repairs had reported that the GCU had effectively been attacked; its engine problem wasn't the result of chance or some manufacturing flaw, it was the result of offensive action.

'Apart from that, and the still unexplained disappearance of an entire star, everything was normal for nearly two decades'. Tishlin's hand flapped once on the table. 'Oh, there were various investigations and boards of inquiry and committees and so on, but the best they could come up with was that the whole thing had been some sort of hi-tech projection, maybe produced by some previously unknown Elder civilisation with a quirky sense of humour, or, even less likely, that the sun and all the rest had popped into Hyperspace and just sped off – though they should have been able to observe that, and hadn't – but basically the whole thing remained a mystery, and after everybody had chewed it over and over till there was nothing but spit left, it just kind of died a natural death.

'Then, over the following seven decades, the *Problem Child* decided it didn't want to be part of Contact any more. It left Contact, then it left the Culture proper and joined the Ulterior – again, very unusual for its class – and meanwhile every single human who'd been on board at the time exercised what are apparently termed Unusual Life Choices'. Tishlin's dubious look indicated he wasn't totally convinced this phrase contributed enormously to the information-carrying capacity of the language. The image made a throat-clearing noise and went on: 'Roughly half of the humans opted for immortality, the other half autoeuthenised. The few remaining humans underwent subtle but exhaustive investigation, though nothing unusual was ever discovered.

'Then there were the ship's drones; they all joined the same Group Mind – again in the Ulterior – and have been incommunicado ever since. Apparently that was even more unusual. Within a century, almost all of those humans who'd opted for immortality were also dead, due to further "semi-contradictory" Unusual Life Choices. Then the Ulterior, and Special Circumstances – who'd taken an interest by this time, not surprisingly – lost touch with the *Problem Child* entirely. It

131

just seemed to disappear, too. The apparition shrugged. 'That was fifteen hundred years ago, Byr. To this day nobody has seen or heard of the ship. Subsequent investigations of the remains of a few of the humans concerned, using improved technology, has thrown up possible discrepancies in the nanostructure of the subjects' brains, but no further investigation has been deemed possible. The story was made public eventually, nearly a century and a half after it all happened; there was even a bit of a media fuss about it at the time, but by then it was a portrait with nobody in it: the ship, the drones, the people; they'd all gone. There was nobody to talk to, nobody to interview, nothing to do profiles of. Everybody was off-stage. And of course the principal celebrities – the star and the artifact – were the most off-stage of all'.

'Well', Genar-Hofoen said. 'All very-'

'Hold on', Tishlin said, holding up one finger. 'There is one loose end. A single traceable survivor from the *Problem Child* who turned up five centuries ago; somebody it might be possible to talk to, despite the fact they've spent the last twenty-four millennia trying to avoid talking'.

'Human?'

'Human', Tishlin confirmed, nodding. 'The woman who was the vessel's formal captain'.

'They still had that sort of thing back then?' Genar-Hofoen said. He smiled. How quaint, he thought.

'It was pretty nominal, even back then', Tishlin conceded. 'More captain of the crew than of the boat. Anyway; she's still around in a sort of abbreviated form'. Tishlin's image paused, watching Genar-Hofoen closely. 'She's in Storage aboard the General Systems Vehicle *Sleeper Service*'.

The representation paused, to let Genar-Hofoen react to the name of the ship. He didn't, not on the outside anyway.

'Just her personality is in there, unfortunately', Tishlin continued. 'Her Stored body was destroyed in an Idiran attack on the Orbital concerned half a millennium ago. I suppose for our purposes that counts as a lucky break; she'd managed to cover her tracks so well – probably with the help of some sympathetic Mind – that if the attack hadn't occurred she'd have remained incognito to this day. It was only when the records were scrutinised carefully after her body's destruction that it was realised who she really was. But the point is that Special Circumstances thinks she might know something about the artifact. In fact, they're sure she does, though it's almost equally certain that she doesn't *know* what she knows'.

Genar-Hofoen was silent for a while, playing with the cord of his dressing gown. The *Sleeper Service*. He hadn't heard that name for a while, hadn't had to think about that old machine for a long time. He'd dreamt about it a few times, had had a nightmare or two about it even, but he'd tried to forget about those, tried to shove those echoes of memories to some distant corner of his mind and been pretty successful at it too, because it felt very strange to be turning over that name in his mind now.

'So why's this all suddenly become important after two and a half millennia?' he asked the hologram.

'Because something with similar characteristics to that artifact has turned up near a star called Esperi, in the Upper Leaf-Swirl, and SC needs all the help it can get to deal with it. There's no trillion-year-old sun-cinder this time, but an apparently identical artifact is just sitting there'.

'And what am I supposed to do?'

'Go aboard the *Sleeper Service* and talk to this woman's Mimage – that's the Mind-stored construct of her personality apparently ...' The image looked puzzled. ... New one on me ... Anyway, you're supposed to try to persuade her to be reborn; talk her into a rebirth so she can be quizzed. The *Sleeper Service* won't just release her, and it certainly won't cooperate with SC, but if she asks to be reborn, it'll let her'.

'But why-?' Genar-Hofoen started to ask.

'There's more', Tishlin said, holding up one hand. 'Even if she won't play, even if she refuses to come back, you're to be equipped with a method of retrieving her through the link you'll forge when you talk to the Mimage, without the GSV knowing. Don't ask me how that's supposed to be accomplished, but I think it's got something to do with the ship they're going to give you to get you to the *Sleeper Service,* after the Affronter ship they're going to hire for you has rendezvoused with it at Tier'.

Genar-Hofoen did his best to look sceptical. 'Is that possible?' he asked. 'Retrieving her like that, I mean. Against the wishes of the *Sleeper*'.

'Apparently', Tishlin said, shrugging. 'SC thinks they've got a way of doing it. But you see what I mean when I said they want you to steal the soul of a dead woman

Genar-Hofoen thought for a moment. 'Do you know what ship this might be? The one to get me to the *Sleeper?'*

'They haven't-' began the image, then paused and looked amused. 'They just told me; it's a GCU called the *Grey Area'*. The image smiled. 'Ah; I see you've heard of it, too'.

'Yeah, I've heard of it', the man said.

The *Grey Area*. The ship that did what the other ships both deplored and despised; actually looked into the minds of other people, using its Electro Magnetic Effectors – in a sense the very, very distant descendants of electronic countermeasures equipment from your average stage three civilisation, and the most sophisticated, powerful but also precisely controllable weaponry the average Culture ship possessed – to burrow into the grisly cellular substrate of an animal consciousness and try to make sense of what it found there for its own – usually vengeful – purposes. A pariah craft; the one the other Minds called *Meatfucker* because of its revolting hobby (though not, as it were, to its face). A ship that still wanted to be part of the Culture proper and nominally still was, but which was shunned by almost all its peers; a virtual outcast amongst the great inclusionary meta-fleet that was Contact.

Genar-Hofoen had heard about the *Grey Area* all right. It was starting to make sense now. If there was one vessel that might be capable of plundering – and, more importantly, that might be willing to plunder – a Stored soul from under the nose of the *Sleeper*, the *Grey Area* was probably it. Assuming what he'd heard about the ship was true, it had spent the last decade perfecting its techniques of teasing dreams and memories out of a variety of animal species, while the *Sleeper Service* had by all accounts been technologically stagnant for the last forty years, its time taken up with the indulgence of its own scarcely less eccentric pastime.

The image of Uncle Tishlin bore a distant expression for a moment, then said, 'Apparently that's part of the beauty of it; just because the *Sleeper Service* is another oddball doesn't mean that it's any more likely than any other GSV to have the *Grey Area* aboard; the GCU will have to lie off, and that'll make this Mimage-stealing trick easier. If the *Grey Area* was actually inside the GSV at the time it probably couldn't carry it off undetected'.

Genar-Hofoen was looking thoughtful again. 'This artifact thing', he said. 'Could almost be a what-do-you-call it, couldn't it? An Outside Context Paradox'.

'Problem', Tishlin said. 'Outside Context Problem'.

'Hmm. Yes. One of those. Almost'.

An Outside Context Problem was the sort of thing most civilisations encountered just once, and which they tended to encounter rather in the same way a sentence encountered a full stop. The usual example given to illustrate an Outside Context Problem was imagining you were a tribe on a largish, fertile island; you'd tamed the land, invented the wheel or writing or whatever, the neighbours were cooperative or enslaved but at any rate peaceful and you were busy raising temples to yourself with all the excess productive capacity you had, you were in a position of near-absolute power and control which your hallowed ancestors could hardly have dreamed of and the whole situation was just running along nicely like a canoe on wet grass ... when suddenly this bristling lump of iron appears sailless and trailing steam in the bay and these guys carrying long funny-looking sticks come ashore and announce you've just been discovered, you're all subjects of the Emperor now, he's keen on presents called *tax* and these bright-eyed holy men would like a word with your priests.

That was an Outside Context Problem; so was the suitably up-teched version that happened to whole planetary civilisations when somebody like the Affront chanced upon them first rather than, say, the Culture.

The Culture had had lots of minor OCPs, problems that could have proved to be terminal if they'd been handled badly, but so far it had survived them all. The Culture's ultimate OCP was popularly supposed to be likely to take the shape of a galaxy-consuming Hegemonising Swarm, an angered Elder civilisation or a sudden, indeed instant visit by neighbours from Andromeda once the expedition finally got there.

In a sense, the Culture lived with genuine OCPs all around it all the time, in the shape of those Sublimed Elder civilisations, but so far it didn't appear to have been significantly checked or controlled by any of them. However, waiting for the first real OCP was the intellectual depressant of choice for those people and Minds in the Culture determined to find the threat of catastrophe even in utopia.

'Almost. Maybe', agreed the apparition. 'Perhaps it's a little less likely to be so with your help'.

Genar-Hofoen nodded, staring at the surface of the table. 'So who's in charge of this?' he asked, grinning. 'There's usually a Mind which acts as incident controller or whatever they call it in something like this'.

'The Incident Coordinator is a GSV called the *Not Invented Here*', Tish told him. 'It wants you to know you can ask whatever you want of it'.

'Uh-huh'. Genar-Hofoen couldn't recall having heard of the ship. 'And why me, particularly?' he asked. He suspected he already had the answer to that one.

'The *Sleeper Service* has been behaving even more oddly than usual', Tishlin said, looking suitably pained. 'It's altered its course schedule, it's no longer accepting people for Storage, and it's almost completely stopped communicating. But it says it will allow you on board'.

'For a brow-beating, no doubt', Genar-Hofoen said, glancing to one side and watching a cloud pass over the meadows of the valley shown on the dining room's projector walls. 'Probably wants to give me a lecture'. He sighed, still looking round the room. He fastened his gaze on Tishlin's simulation again. 'She still there?' he asked.

The image nodded slowly.

'Shit', Genar-Hofoen said.

III

'But it makes my brain hurt'.

'Nevertheless, Major. This is of inestimable importance'.

'I only looked at the first bit there and it's already given me a thumping case-ache'.

'Still, it has to be done. Kindly read it all carefully and then I'll explain its significance'.

'Knot my stalks, this is a terrible thing to ask of a chap after a regimental dinner'. Fivetide wondered if humans suffered so for their self-indulgence. He doubted it, no matter what they claimed; with the possibly honourable, possibly demented exception of Genar-Hofoen, they seemed a bit too stuffy and sensible willingly to submit to such self-punishment in the cause of fun. Besides, they were so insecure in their physical inheritance they had meddled with themselves in all sorts of ways; probably they thought hangovers were just annoying, rather than character-forming and so had, shortsightedly, dispensed with them.

'I realise it's early and it is the morning after the night before, Major. But please'.

The emissary – which Fivetide had met once before, and which possessed the irritating trait of looking somewhat like a better-built version of Fivetide's dear departed father – had just appeared in the nest house without notice or warning. If he hadn't known the way these things worked, Fivetide would right now be thinking of ways to torture the head of nest security. Tentacles had rolled, beaks had been separated, for less.

Lucky he'd been able to whip the bed covers round his deputy wife and both vice courtesans before the blighter had announced his/its presence by just floating into the nest.

Fivetide clapped his forebeak together a couple of times. *Tastes like I've had me beak up me arse*, he thought. 'Can't you just tell me what the damn signal means now?' he asked.

'You won't know what I'm referring to. Come now; the sooner you read it the sooner I'll be able to tell you what it means, and the sooner I'll be able to demonstrate how it is just possible that this information will – at the very least – enable you to remove the harness of Culture interference forever'.

'Hmm. I'm sure. And what'll it do at most?'

The emissary of the ship let its eye stalks dip to either side, the Affronter equivalent of a smile. 'At most, the information in this signal will lead to you being able to dominate the Culture as completely as it – if it chose to – could dominate you'. The creature paused. 'This signal could conceivably presage the start of a process which will deliver the entire Galaxy into your hands, and subsequently open up territories for expansion and exploitation beyond that which you cannot even begin to guess at. And I do *not* exaggerate. Have I your attention now, Major?'

Fivetide snorted sceptically. 'I suppose you have', he said, shaking his limbs and rubbing his eyes. He returned his gaze to the note screen, and read the signal.

xGCU *Fate Amenable To Change,*
    oGSV *Ethics Gradient*
    & strictly as SC cleared:
    Excession notice @c18519938.52314.
    Constitutes formal All-ships Warning Level 0
    [(in temporary sequestration) - textual note added by GSV Wisdom Like Silence @ n4.28.855.0150.650001].
Excession.
Confirmed precedent-breach. Type K7^. True class non-estimal. Its status: Active. Aware. Contactiphile. Uninvasive sf. LocStatre: Esperi (star).
First ComAtt (its, following shear-by contact via my primary scanner @ n4.28.855.0065.59312) @ n4.28.855.0065.59487 in M1-a16 & Galin II by tight beam, type 4A. PTA & Handshake burst as appended, x@ O.7Y. Suspect signal gleaned from Z-E/Ialsaer ComBeam spread, 2nd Era. xContact cailsigned 'I'. No other signals registered.

My subsequent actions: maintained course and speed, skim-declutched primary scanner to mimic 50% closer approach, began directed full passive HS scan (sync./start of signal sequence, as above), sent buffered Galin II proforma message-reception confirmation signal to contact location, dedicated track scanner @ 19% power and 300% beamspread to contact @ –5% primary scanner roll-off point, Instigated ^{2}exponential slow-to-stop line manoeuvre synchronised to skein-local stop-point @ 12% of track scanner range limit, ran full systems check as detailed, executed slow/4 swing-around then retraced course to previous closest

approach point and stop @ standard ^{2}ex curve. Holding there.
Excession's physical characteristics: (!am i) sphere rad. 53.34km, mass (non-estimal by space-time fabric influence – locality ambiently planar – estimated by pan-polarity material density norms at) 1.45×8^{13}. Layered fractal matter-type-intricate structure, self supporting, open to (field-filtered) vacuum, anomalous field presence inferred from 8^{21} kHz leakage. Affirm K7^ category by HS topology & eG links (inf. & ult.). eG link details non-estimal. DiaGlyph files attached.
Associated anomalous materials presence: several highly dispersed detritus clouds all within 28 minutes, three consistent with staged destruction of >.1 m^3 near-equiv-tech entity, another ditto approx 3^8 partially exhausted M-DAWS .1 cal rounds, another consisting of general hi-soph level (O_2-atmosphered) ship-internal combat debris. Latter drifting directly away from excession's current position. Retracks of debris clouds' expansion profiles indicates mutual age of 52.5 days. Combat debris cloud Implicitly originating @ a point 948 milliseconds from excession's current position. DiaGlyph files attached.
No other presences apparent to within 30 years.
My status: H&H, unTouched. L-8 secure post system-scour (100%). ATDPSs engaged. CRTTDPSs engaged.
Repeat:
Excession eG (inf. & ult) linked, confirmed.
eGrid link details non-estimal.
True class non-estimal.
Awaiting.
@ n4.28.855.0073.64523...
... PS:
Gulp.

Fivetide shook his stalks. Gods, this hangover was fierce.
'All right', he said, 'I've read, but I still don't understand'.
The emissary of the war vessel *Attitude Adjuster* smiled again.
'Allow me to explain'.

SCIENCE AND THE RETREAT FROM REASON

JOHN GILLOTT

John Gillott's background is in mathematics. He has researched and written on a wide range of environmental and scientific issues. Currently he works on policy at the Genetic Interest Group, London. With Manjit Kumar he is author of Science and the Retreat from Reason. *Together they are currently working on a book looking at the dynamics of the debate between the defenders and critics of modern science.*

Do we live in a scientific or an anti-scientific age? When it is posed like this, as the debate on the attitude of contemporary culture toward science often is, a number of striking paradoxes emerge.

Take, for a start, the indisputable evidence of powerful mystical, anti-scientific prejudices in society. Many prominent scientists and commentators have bemoaned the popularity of astrology and 'UFOlogy' (not forgetting television programmes such as *The X-Files*). Others have challenged these trends in academia. In their contentious work, *Higher Superstition: The Academic Left and Its Quarrels With Science*, Paul Gross and Norman Levitt take issue with the growing body of work produced by relativistic sociologists of science. In his *Einstein, History and Other Passions*, Gerald Holton, the eminent historian of science, cites the recent critical exhibition on the history of science mounted by the Smithsonian Institution – "Science in American Life" – as an illustration of intellectual disillusionment with science.

Athough I share many of these concerns, I do not believe that it is accurate simply to characterise our times as 'anti-science'. Jon Miller, vice president of the Chicago Academy of Sciences, points out that while 'many scientists believe there is an anti-science attitude' among the public, his survey of public attitudes suggests the opposite: 'Americans believe in science almost as an article of faith.' In its 1995 survey of attitudes toward science, the *Daily Telegraph* highlighted the apparent enthusiasm of the British public for science and its applications: 54% of those questioned wanted the Government's Millennium Project to make healthcare a priority. The most popular choice for 'more press coverage' (with 34% support) was medical discoveries.

For the editorial team on the *Public Understanding of Science* journal based at Britain's Science Museum, opinion surveys indicate that commentators who complain of pervasive anti-science trends exaggerate the scale of the problem. Indeed, among the young in particular, polls reveal an enthusiasm for science. Proponents of this 'don't panic' school of thought point to the popularity of the film *Jurassic Park* – the kids loved it, they say. But this is a dubious argument. The central message of *Jurassic Park* is that human meddling in nature has potentially catastrophic consequences. Director Steven Spielberg candidly declared that his objective in making the film was to 'bring science down a peg or two'.

Anti-science trends in a broader context

In an important respect we do live in an irrational, anti-scientific age. But to be more precise, an age that is very dependent upon science, even welcoming of science in a restricted sense, but fundamentally uneasy with it because of a deep concern about broader projects linked to science. Anti-science trends do exist and are influential, but it is too simplistic to say the contemporary mood is either for or against science. The real significance of anti-science moods is that they are only an aspect of or a consequence of something else: a strongly-rooted loss of faith in human social advance.

One important element of the broader attitudes which influence attitudes toward science is the widely-held view that human progress based upon an interventionist, manipulative relationship to the natural world is neither feasible nor desirable. Theorists of the 'Risk Society' condemn the project of modernity, which they identify as the attempt to control nature for human ends, leading to a proliferation of risks which in turn undermine the modern project itself. They warn of the impending 'revenge of nature' against human arrogance, and suggest that the dangers unleashed by scientific and social progress outweigh the benefits.

A distinct lack of confidence in humanity's creative abilities is widespread today. Indeed many of the broader visions of the future in today's popular culture denigrate and belittle humanity. Malcolm Gladwell's observation that popular science writing on the spectre of lethal epidemics of new infections exhibits nothing less than a self-loathing of humanity is accurate and has a broader relevance. Today, human action is associated more with destruction than with creativity – and this reflects back on science on account of the power science hands to humanity.

Attitudes to genetic science provide a good illustration of the point that society's worries with science can only be understood in relation to a broader rejection of what can be loosely termed 'progress'. Surveys show that there is widespread support for genetic science as a means to tackle disease. Indeed, many scientists in the field are concerned that the public discussion of genetics may have raised expectations far beyond what is likely to be possible in the near future. But enthusiasm rapidly evaporates once genetics is associated with anything other than repairing what can be thought of as nature's mistakes. Beyond repair work we rapidly hit talk of the dangers of 'designer babies'; a Brave New World in which the Spectre of Eugenics hovers over the field. Such is the level of distrust of human design in this area that the idea of using genetic manipulation to enhance resistance to disease is widely rejected. It is this distrust of human attempts at design – a distrust that, as Steven Spielberg recognised, generates a suspicion of science – that is the central theme of *Jurassic Park*.

We can develop this point further by considering the following three well-known theses on contemporary attitudes toward science.

Blaming the critics
According to Paul Gross and Norman Levitt, the growth of relativism is not only a symptom of the growth of anti-science trends, but is also a primary cause of them. As Daniel Kleppner of MIT put it at a conference called the 'Flight from Science and Reason' organised by Gross and Levitt in 1995, the prejudice against science is a part of the whole post-modern "shtick".

Traditionalists
For Lewis Wolpert and John Maddox, it is that fact that science is difficult, counter-intuitive, even an unnatural way of thinking that explains suspicion of science. That it is also a challenge to cherished beliefs and is in parts bleak in its message only compounds the problem.

Sociologists
According to Anthony Giddens, science is an 'expert system': we have to *trust* science without understanding it. But the inescapable obverse of trust is doubt: 'ignorance always provides grounds for scepticism or at least caution'. John Durant, Britain's first professor in the public understanding of science, emphasises that it is the contemporary 'pervasiveness and power' of science that explains the demise of 'the rosy optimism of the Victorian age'.

Although many of the criticisms made by Gross and Levitt of the absurdities of 'perspectivism' in science are sound, their attempt to jump from these issues to an understanding of contemporary anti-science trends fails. The thesis is circular: suspicion of science is said to be generated by people who are suspicious of it. But, why do anti-science moods have such a wide influence? And why now? If it is all the fault of the relativist left, then they have a lot more influence than anybody has noticed.

Wolpert and Maddox are right, science is both counter-intuitive and difficult. But why should this generate suspicion? It is quite possible to appreciate science without understanding it in depth. And why now? Science has always been difficult, but it has only recently become unpopular. Durant's adaptation of Giddens shares the ahistorical character of the traditionalists' defence. Science was a pervasive expert system 30 years ago, yet it enjoyed much higher prestige in society then than it does today.

To explain the distinctive unpopularity of science today we need to take a wide view of the place of science in society, and in particular to consider the declining appeal of any concept of social progress. The rising influence of environmental politics reflects this trend. As Anna Bramwell has observed, in the course of the 1960s, among young people in particular, a remarkable reversal in attitudes took place. Before the 1960s, 'Nature' was generally regarded as a negative influence on human society, as a force to be overcome through science, technology and social organisation. But since the 1960s, 'Nature' has come to be seen as a beneficent, if not transcendent, agency, and indeed one increasingly in jeopardy as a result of human intervention. This dramatic shift has inevitably altered society's perspective on science. Though the practical benefits of science are acknowledged, it is also distrusted, since every significant scientific advance only increases humanity's power to alter nature – and this is thought to be a source of environmental deterioration and human inequality.

The sociologists have a point, society is ambivalent about science. But this is not so much the result of suspicion of 'expert systems', as an expression of insecurity about the ability of humanity to carry through interventionist projects without throwing up problems that it is incapable of solving. Take, for example, the writings of US vice-president Al Gore. As the 'technology czar' of the Clinton administration, he is keen to promote science and he is surely just the kind of person who ought to feel comfortable with 'expert systems'. But Gore is also the man who wrote the doom-laden tract *Earth in the Balance* in the years immediately before the Democratic party's capture of the White House. Arguing that modern science has done much to destroy the environment and erode the values of American society, Gore warned that many scientists are playing at God, 'unaccompanied by godlike wisdom'. He is a promoter of science only in so far as science is divorced from any connection to social progress.

In summary, while scientists like Gross and Levitt, Wolpert and Maddox are to be praised for upholding objectivity and rationality against the 57 brands of relativism, they lack sufficient sociological sensitivity to grasp the specific character of the contemporary debate about science. But given that influential sociologists also miss the mark, this is perhaps not surprising.

The retreat from reason – a problem for both society and science

In our book, *Science and the Retreat from Reason*, Manjit Kumar and myself focus on the key paradox of our era: the fact that science continues to advance while society – and even scientists themselves –

retreats from rationality and experimentation. The coexistence of these two trends is crucial for understanding contemporary suspicions of science. Indeed, the influence of anti-science trends alongside anti-progress trends in society today is largely a consequence of the fact that scientific advance raises awkward questions for a society fearful of change. While science and social progress are distinct, our analysis locates contemporary anti-science trends as an integral part of a wider rejection of 'progress'. At the same time, the denigration of science inevitably also belittles humanity.

The fashionable concepts of the 'Precautionary Principle' and the 'Risk Society' provide good illustrations of the links between a misanthropic social outlook and a negative attitude toward science. When Brian Wynne and Sue Mayer called for a "greener" culture of good science', Alex Milne rightly rejected the idea on the grounds that what was being proposed amounted to moral philosophy not science (*New Scientist*, 5 and 12 June 1993). But there is more to be said than this, for the real content of much of the writing on risk and science is moral philosophy of a particular kind, one that is hostile to interventionist attitudes toward nature, and motivated by a desire to belittle science.

So much for diagnosis. I now want to move on to articulate a critique of contemporary hostility to progress and science. This is vital, not only because I believe that many of the specific claims made by the critics of science are wrong, but also because they are harmful to both human interests and science.

The panic over Mad Cow disease (BSE) in Britain in spring 1996 provoked considerable controversy about questions of science and society. Oxford philosopher John Gray captured the mood well in an article in *The Guardian* (26 March 1996) which drew links between BSE and the accident at Chernobyl. As a result of the 'culture of technological mastery of nature', we face, according to Gray, 'incalculable but catastrophic' risks. All in all, he asked rhetorically, 'is it altogether fanciful to see the threat of a major outbreak of CJD as a symptom of nature's rebellion against human hubris?'.

Speaking at a conference in Britain on the Politics of Risk Society at the height of the BSE controversy, the German sociologist Ulrich Beck echoed Gray's concerns. Adding genetic engineering to BSE and Chernobyl, he warned that 'we are in danger of creating a situation where alarmingly large risks are nobody's responsibility'. 'Neglecting risks', he suggested, 'is one of the most effective ways of reinforcing them' (*The Independent*, 26 March 1996). Beck condemned scientists for conducting live experiments on society in ignorance of the consequences.

For Gray the threat of a "revenge of nature" rises in proportion to the level of technology used, which increases with the distance from natural processes. Thus modern farming methods carry greater risks than traditional ones because they are unnatural; nuclear power is especially dangerous; and genetic engineering 'must be viewed with suspicion' because we are unaware of the longer term consequences of meddling. If we are to learn the lesson of Chernobyl and BSE, he says, then "we should be ready to err on the side of caution".

The call to 'play safe' has an undoubted appeal in our insecure and anxious society. It is popular among those who are suspicious of the motives of commercial organisations involved in genetics and the nuclear industry. It connects well with concerns about environmental pollution. And it taps directly into a fear of the unknown. But in all cases there is an exaggeration of risks. Nuclear power has led to the loss of far fewer lives than other forms of energy production; 10 years after the release of the first genetically engineered organism we are still waiting for the disaster predicted by environmentalists; and CJD is likely to remain a relatively rare disease, despite the probable link between BSE and new-variant CJD.

In a society which always fears the worst, the inevitable uncertainty associated with scientific innovation can readily lend legitimacy to the doomsday scenario. Beck has built a whole sociological outlook out of celebrating uncertainty and inflating risk, which culminates in a questioning of the usefulness of social and scientific progress, as well as the efficacy of human action in the face of change. Uncertainty is a fact of life in science as in everything else. Beck's trick is to use this to legitimate insecurity about human judgment and action.

For Beck, the themes of the 'Risk Society' rest upon the possibility of the ultimate worst-case scenario – human extinction: 'I use the term "risk society"', he writes in his book *Ecological Politics in an Age of Risks*, 'for those societies that are confronted by the challenges of the self-created possibility, hidden at first, then increasingly apparent, of the self-destruction of all life on this earth.' But just what might bring this about? Well, in Beck's dystopic vision of the world, almost anything. He proceeds from the abstract postulation of the possibility of extinction to speculate in a fearful and gloomy tone about the potential of global Armageddon resulting from familiar threats to the environment.

The development of the 'revenge of nature' thesis to apocalyptic images of extinction has many harmful consequences. For a start this outlook belittles human knowledge and underestimates the scope for interventions to contain environmental dangers. But for many commentators, this is not a problem. Indeed, for some this is a specific political objective: they would like to bring science down a peg or two. This is a principal theme running through *Misunderstanding Science? The Public Reconstruction of Science and Technology*, a collection edited by Alan Irwin and Brian Wynne, which seeks to bring the sociological perspective of Beck and others to bear upon science and technology. In their conclusion, Irwin and Wynne call for scientists to be more 'humble' before the court of public opinion and tell them to stop claiming access to a 'higher rationality'. But what is the justification of science if it is not superior to other forms of knowledge? Of course the findings of scientists can and should be challenged – but a challenge must be based on a critical study of the evidence; that is, findings should be contested in a scientific manner. By contrast, Irwin and Wynne, in the name of 'contextualising' science, legitimate its degradation.

Risk theory has another harmful consequence. One the one hand, its proponents insist that risks are uncertain because the consequences of scientific intervention cannot be predicted. Yet on the other hand they tell scientists to restrain their experimental urges until they can quantify risks. This can only mean that novel technologies and procedures should not be tested since the attendant risks cannot be foreseen. The logic of this argument would lead to the suspension of all experimentation and innovation, and the establishment of caution as the prime determinant of all human action.

It is easy to feel swamped by problems if we don't try to overcome them. The fact is that there are risks in new procedures, and there are unforeseen consequences. But this is no bad thing, for we make progress by overcoming problems. And, you never know, the unforeseen consequences might turn out to be beneficial. 'Live experiments' have been the source of new knowledge and social gains throughout the ages. The 'revenge of nature' idea belittles past achievements and current abilities to deal with problems. The call for caution in the face of risks does the same and more, since it also robs us of the chance of proving our abilities in practice.

John Gray at least acknowledges that 'everything that is worthwhile' in modern society results from the fact that we are no longer dominated by the vicissitudes of nature. But for Gray this perspective only holds in relation to the past; it has no relevance to the present and the future because he has given up on the possibility of taking society forward. He concludes his recent book *Enlightenment's Wake* with a rejection of even the possibility of collective human advance. Like many contemporary environmentalist and feminist writers on science he argues that science and its aspirations to control are products of human hubris that are best abandoned.

Conclusion

The two central themes of contemporary writings on science – the desire to bring science down a peg or two and the broader anti-experimental attitude – are closely linked. One direct link is that belittling science also amounts to a belittling of humanity. This is not just because science is a great achievement of human thought and action, but also because science deserves to occupy a special place in society. Why? Because science offers us the chance to improve our lot. It is a mark of civilization that we have been able to use our knowledge of a natural world which is not uniformly a benign force in part to shelter ourselves from its excesses. To belittle science is therefore to belittle civilization.

Science does not equal social progress. It is, however, an important tool for any society that wants to advance. For those concerned to develop a proper appreciation of the role and potential of science, the lesson to be drawn from the contemporary discussion of science and social progress is that we will not generate a better appreciation of the role of science in society by simply championing science. We must uphold and defend science against its critics, but we must do this as an element of a broader challenge to the sense of caution and conservatism which grips society today.

THE CHEMISTRY LESSON
COMBUSTION

The bunsen gasses into flame.
(Note: the inner cone of blue will not burn.)
Other things do – iron filings spray from the jet;
copper glows itself into a green dream, imaging
another, better world than this;
calcium is house-brick red; magnesium
flares and fetters my eyes to the wall;
aluminium flashes; ammonium dichromate,
pestle-crushed, dies in a volcano.

We smell the devil, we see his sulphur
bleach a tulip white. Phosphorus
(beware of burns) emerges from water
and sets itself on flame. (Return to flask.)
We burn our food: butter, off the sandwich,
will ignite and give us gold, its luminous
butter-flame; the bread burns black;
alcohol (a food?) is a clean burn –
no carbonizing, no traces of earth
in its release.

Fabrics, too, will take to the flame:
cotton, a gracious, slow-living fire;
wool – prickly, short-tempered, but,
as ash, submissive and damp; silk?
as you might expect, it – she – lifts
her own skirts away from the flame,
and when at last it does ignite,
it has the freshness of a clear morning's
sun, bright and resilient.

What fails to burn are flowers, being
already the flame the plant has made.
Dwindling and going to vapour, the petals
have their place in our mind's eye,
which shares in their unwillingness to perform
solely for the delight of substance.

Ranald Macdonald

WHAT MATHEMATICS IS FOR

IAN STEWART

We've now established the uncontroversial idea that nature is full of patterns. But what do we want to do with them? One thing we can do is sit back and admire them. Communing with nature does all of us good: it reminds us of what we are. Painting pictures, sculpting sculptures, and writing poems are valid and important ways to express our feelings about the world and about ourselves. The entrepreneur's instinct is to exploit the natural world. The engineer's instinct is to change it. The scientist's instinct is to try to understand it – to work out what's really going on. The mathematician's instinct is to structure that process of understanding by seeking generalities that cut across the obvious subdivisions. There is a little of all these instincts in all of us, and there is both good and bad in each instinct.

I want to show you what the mathematical instinct has done for human understanding, but first I want to touch upon the role of mathematics in human culture. Before you buy something, you usually have a fairly clear idea of what you want to do with it. If it is a freezer, then of course you want it to preserve food, but your thoughts go well beyond that. How much food will you need to store? Where will the freezer have to fit? It is not always a matter of utility; you may be thinking of buying a painting. You still ask yourself where you are going to put it, and whether the aesthetic appeal is worth the asking price. It is the same with mathematics – and any other intellectual worldview, be it scientific, political, or religious. Before you buy something, it is wise to decide what you want it for.

So what do we want to get out of mathematics?

Each of nature's patterns is a puzzle, nearly always a deep one. Mathematics is brilliant at helping us to solve puzzles. It is a more or less systematic way of digging out the rules and structures that lie behind some observed pattern or regularity, and then using those rules and structures to explain what's going on. Indeed, mathematics has developed alongside our understanding of nature, each reinforcing the other. I've mentioned Kepler's analysis of snowflakes, but his most famous discovery is the shape of planetary orbits. By performing a mathematical analysis of astronomical observations made by the contemporary Danish astronomer Tycho Brahe, Kepler was eventually driven to the conclusion that planets move in ellipses. The ellipse is an oval curve that was much studied by the ancient Greek geometers, but the ancient astronomers had preferred to use circles, or systems of circles, to describe orbits, so Kepler's scheme was a radical one at that time.

People interpret new discoveries in terms of what is important to them. The message astronomers received when they heard about Kepler's new idea was that neglected ideas from Greek geometry could help them solve the puzzle of predicting planetary motion. It took very little imagination for them to see that Kepler had made a huge step forward. All sorts of astronomical phenomena, such as eclipses, meteor showers, and comets, might yield to the same kind of mathematics. The message to mathematicians was quite different. It was that ellipses are really interesting curves. It took very little imagination for them to see that a general theory of curves would be even more interesting. Mathematicians could take the geometric rules that lead to ellipses and modify them to see what other kinds of curve resulted.

Similarly, when Isaac Newton made the epic discovery that the motion of an object is described by a mathematical relation between the forces that act on the body and the acceleration it experiences, mathematicians and physicists learned quite different lessons. However, before I can tell you what these lessons were I need to explain about acceleration. Acceleration is a subtle concept: it is not a fundamental quantity, such as length or mass; it is a rate of change. In fact, it is a "second order" rate of change – that is, a rate of change of a rate of change. The velocity of a body – the speed with which it moves in a given direction – is just a rate of change: it is the rate at which the body's distance from some chosen point changes. If a car moves at a steady speed of sixty miles per hour, its distance from its starting point changes by sixty miles every hour. Acceleration is the rate of change of velocity. If the car's velocity increases from sixty miles per hour to sixty-five miles per hour, it has accelerated by a definite amount. That amount depends not only on the initial and final speeds, but on how quickly the change takes place. If it takes an hour for the car to increase its speed by five miles per hour, the acceleration is very small; if it takes only ten seconds, the acceleration is much greater.

I don't want to go into the measurement of accelerations. My point here is more general: that acceleration is a rate of change of a rate of change. You can work out distances with a tape measure, but it is far harder to work out a rate of change of a rate of change of distance. This is why it took humanity a long time, and the genius of a Newton, to discover the law of motion. If the pattern had been an obvious feature of distances, we would have pinned motion down a lot earlier in our history.

In order to handle questions about rates of change, Newton – and independently the German mathematician Gottfried Leibniz – invented a new branch of mathematics, the calculus. It changed the face of the Earth – literally and metaphorically. But, again, the ideas sparked by this discovery were different for different people. The physicists went off looking for other laws of nature that could explain natural phenomena in terms of rates of change. They found them by the bucketful – heat, sound, light, fluid dynamics, elasticity, electricity, magnetism. The most esoteric modern theories of fundamental particles still use the same general kind of mathematics, though the interpretation – and to some extent the implicit worldview – is different. Be that as it may, the mathematicians found a totally different set of questions to ask. First of all, they spent a long time grappling with what "rate of change" really means. In order to work out the velocity of a moving object, you must measure where it is, find out where it moves to a very short interval of time later, and divide the distance moved by the time elapsed. However, if the body is accelerating, the result depends on the interval of time you use. Both the mathematicians and the physicists had the same intuition about how to deal with this problem: the interval of time you use should be as small as possible. Everything would be wonderful if you could just use an interval of zero, but unfortunately that won't work, because both the distance traveled and the time elapsed will be zero, and a rate of change of $0/0$ is meaningless. The main problem with nonzero intervals is that whichever one you choose, there is always a smaller one that you could use instead to get a more accurate answer. What you would really like is to use the smallest possible nonzero interval of time – but there is no such thing, because given any nonzero number, the number half that size is also nonzero. Everything would work out fine if the interval could be made infinitely small – "infinitesimal." Unfortunately, there are difficult logical paradoxes associated with the idea of an infinitesimal; in particular, if we restrict ourselves to numbers in the usual sense of the word, there is no such thing. So for about two hundred years, humanity was in a very curious position as regards the calculus. The physicists were using it, with great success, to understand nature and to predict the way nature behaves; the mathematicians were worrying about what it really meant and how best to set it up so that it worked as a sound mathematical theory; and the philosophers were arguing that it was all nonsense. Everything got resolved eventually, but you can still find strong differences in attitude.

The story of calculus brings out two of the main things that mathematics is for: providing tools that let scientists calculate what nature is doing, and providing new questions for mathematicians to sort out to their own satisfaction. These are the external and internal aspects of mathematics, often referred to as applied and pure mathematics (I dislike both adjectives, and I dislike the implied separation even more). It might appear in this case that the physicists set the agenda: if the methods of calculus seem to be working, what does it matter *why* they work? You will hear the same sentiments expressed today by people who pride themselves on being pragmatists. I have no difficulty with the proposition that in many respects they are right. Engineers designing a bridge are entitled to use standard mathematical methods even if they don't know the detailed and often esoteric reasoning that justifies these methods. But I, for one, would feel uncomfortable driving across that bridge if I was aware that *nobody* knew what justified those methods. So, on a cultural level, it pays to have some people who worry about pragmatic methods and try to find out what really makes them tick. And that's one of the jobs that mathematicians do. They enjoy it, and the rest of humanity benefits from various kinds of spin-off, as we'll see.

In the short term, it made very little difference whether mathematicians were satisfied about the logical soundness of the calculus. But in the long run the new ideas that mathematicians got by worrying about these internal difficulties turned out to be very useful indeed to the outside world. In Newton's time, it was impossible to predict just what those uses would be, but I think you could have predicted, even then, that uses would arise. One of the strangest features of the relationship between mathematics and the "real world," but also one of the strongest, is that good mathematics, *whatever its source*, eventually turns out to be useful. There are all sorts of theories why this should be so, ranging from the structure of the human mind to the idea that the universe is somehow built from little bits of mathematics. My feeling is that the answer is probably quite simple: mathematics is the science of patterns, and nature exploits just about every pattern that there is. I admit that I find it much harder to offer a convincing reason for nature to behave in this manner. Maybe the question is back to front: maybe the point is that creatures able to ask that kind of question can evolve only in a universe with that kind of structure.*

Whatever the reasons, mathematics definitely is a useful way to think about nature. What do we want it to tell us about the patterns we observe? There are many answers. We want to understand how they happen; to understand *why* they happen, which is different; to organize the underlying patterns and regularities in the most satisfying way; to predict how nature will behave; to control nature for our own ends; and to make practical use of what we have learned about our world. Mathematics helps us to do all these things, and often it is indispensable.

*This explanation, and others, are discussed in *The Collapse of Chaos*, by Jack Cohen and Ian Stewart (New York: Viking, 1994)

For example, consider the spiral form of a snail shell. *How* the snail makes its shell is largely a matter of chemistry and genetics. Without going into fine points, the snail's genes include recipes for making particular chemicals and instructions for where they should go. Here mathematics lets us do the molecular bookkeeping that makes sense of the different chemical reactions that go on; it describes the atomic structure of the molecules used in shells, it describes the strength and rigidity of shell material as compared to the weakness and pliability of the snail's body, and so on. Indeed, without mathematics we would never have convinced ourselves that matter really is made from atoms, or have worked out how the atoms are arranged. The discovery of genes – and later of the molecular structure of DNA, the genetic material – relied heavily on the existence of mathematical clues. The monk Gregor Mendel noticed tidy numerical relationships in how the proportions of plants with different characters, such as seed color, changed when the plants were crossbred. This led to the basic idea of genetics – that within every organism is some cryptic combination of factors that determines many features of its body plan, and that these factors are somehow shuffled and recombined when passing from parents to offspring. Many different pieces of mathematics were involved in the discovery that DNA has the celebrated double-helical structure. They were as simple as Chargaff's rules: the observation by the Austrian-born biochemist Erwin Chargaff that the four bases of the DNA molecule occur in related proportions; and they are as subtle as the laws of diffraction, which were used to deduce molecular structure from X-ray pictures of DNA crystals.

The question of *why* snails have spiral shells has a very different character. It can be asked in several contexts – in the short-term context of biological development, say, or the long-term context of evolution. The main mathematical feature of the developmental story is the general shape of the spiral. Basically, the developmental story is about the geometry of a creature that behaves in much the same way all the time, but keeps getting bigger. Imagine a tiny animal, with a tiny protoshell attached to it. Then the animal starts to grow. It can grow most easily in the direction along which the open rim of the shell points, because the shell gets in its way if it tries to grow in any other direction. But, having grown a bit, it needs to extend its shell as well, for self-protection. So, of course, the shell grows an extra ring of material around its rim. As this process continues, the animal is getting bigger, so the size of the rim grows. The simplest result is a conical shell, such as you find on a limpet. But if the whole system starts with a bit of a twist, as is quite likely, then the growing edge of the shell rotates slowly as well as expanding, and it rotates in an off-centered manner. The result is a cone that twists in an ever-expanding spiral. We can use mathematics to relate the resulting geometry to all the different variables – such as growth rate and eccentricity of growth – that are involved. If, instead, we seek an evolutionary explanation, then we might focus more on the strength of the shell, which conveys an evolutionary advantage, and try to calculate whether a long thin cone is stronger or weaker than a tightly coiled spiral. Or we might be more ambitious and develop mathematical models of the evolutionary process itself, with its combination of random genetic change – that is, mutations – and natural selection.

A remarkable example of this kind of thinking is a computer simulation of the evolution of the eye by Daniel Nilsson and Susanne Pelger, published in 1994. Recall that conventional evolutionary theory sees changes in animal form as being the result of random mutations followed by subsequent selection of those individuals most able to survive and reproduce their kind. When Charles Darwin announced this theory, one of the first objections raised was that complex structures (like an eye) have to evolve fully formed or else they won't work properly (half an eye is no use at all), but the chance that random mutation will produce a coherent set of complex changes is negligible. Evolutionary theorists quickly responded that while half an eye may not be much use, a half-*developed* eye might well be. One with a retina but no lens, say, will still collect light and thereby detect movement; and any way to improve the detection of predators offers an evolutionary advantage to any creature that possesses it. What we have here is a verbal objection to the theory countered by a verbal argument. But the recent computer analysis goes much further.

It starts with a mathematical model of a flat region of cells, and permits various types of "mutation." Some cells may become more sensitive to light, for example, and the shape of the region of cells may bend. The mathematical model is set up as a computer program that makes tiny random changes of this kind, calculates how good the resulting structure is at detecting light and resolving the patterns that it "sees" and selects any changes that improve these abilities. During a simulation that corresponds to a period of about four hundred thousand years – the blink of an eye, in evolutionary terms – the region of cells folds itself up into a deep, spherical cavity with a tiny irislike opening and, most dramatically, a lens. Moreover, like the lenses in our own eyes, it is a lens whose refractive index – the amount by which it bends light – varies from place to place. In fact, the pattern of variation of refractive index that is produced in the computer simulation is very like our own. So here mathematics shows that eyes definitely can evolve gradually and naturally, offering increased survival value at every stage. More than that: Nilsson and Pelger's work demonstrates that given certain key biological faculties (such as cellular receptivity to light, and cellular mobility), structures remarkably similar to eyes *will* form – all in line with Darwin's principle of natural selection. The mathematical model provides a lot of extra detail that the verbal Darwinian argument can only guess at, and gives us far greater confidence that the line of argument is correct.

I said that another function of mathematics is to organize the underlying patterns and regularities in the most satisfying way. To illustrate this aspect, let me return to the question raised in the first chapter. Which – if either – is significant: the three-in-a-row pattern of stars in Orion's belt, or the three-in-a-row pattern to the periods of revolution of Jupiter's satellites? Orion first. Ancient human civilizations organized the stars in the sky in terms of pictures of animals and mythic heroes. In these terms, the alignment of the three stars in Orion appears significant, for otherwise the hero would have no belt from which to hang his sword. However, if we use three-dimensional geometry as an organizing principle and place the three stars in their correct positions in the heavens, then we find that they are at very different distances from the Earth. Their equispaced alignment is an accident, depending on the position from which they are being viewed. Indeed, the very word "constellation" is a misnomer for an arbitrary accident of viewpoint.

The numerical relation between the periods of revolution of Io, Europa, and Ganymede could also be an accident of viewpoint. How can we be sure that "period of revolution" has any significant meaning for nature? However, that numerical relation fits into a dynamical framework in a very significant manner indeed. It is an example of a *resonance*, which is a relationship between periodically moving bodies in which their cycles are locked together, so that they take up the same relative positions at regular intervals. This common cycle time is called the period of the system. The individual bodies may have different – but related – periods. We can work out what this relationship is. When a resonance occurs, all of the participating bodies must return to a standard reference position after a whole number of cycles – but that number can be different for each. So there is some common period for the system, and therefore each individual body has a period that is some whole-number divisor of the common period. In this case, the common period is that of Ganymede, 7.16 days. The period of Europa is very close to half that of Ganymede, and that of Io is close to one-quarter. Io revolves four times around Jupiter while Europa revolves twice and Ganymede

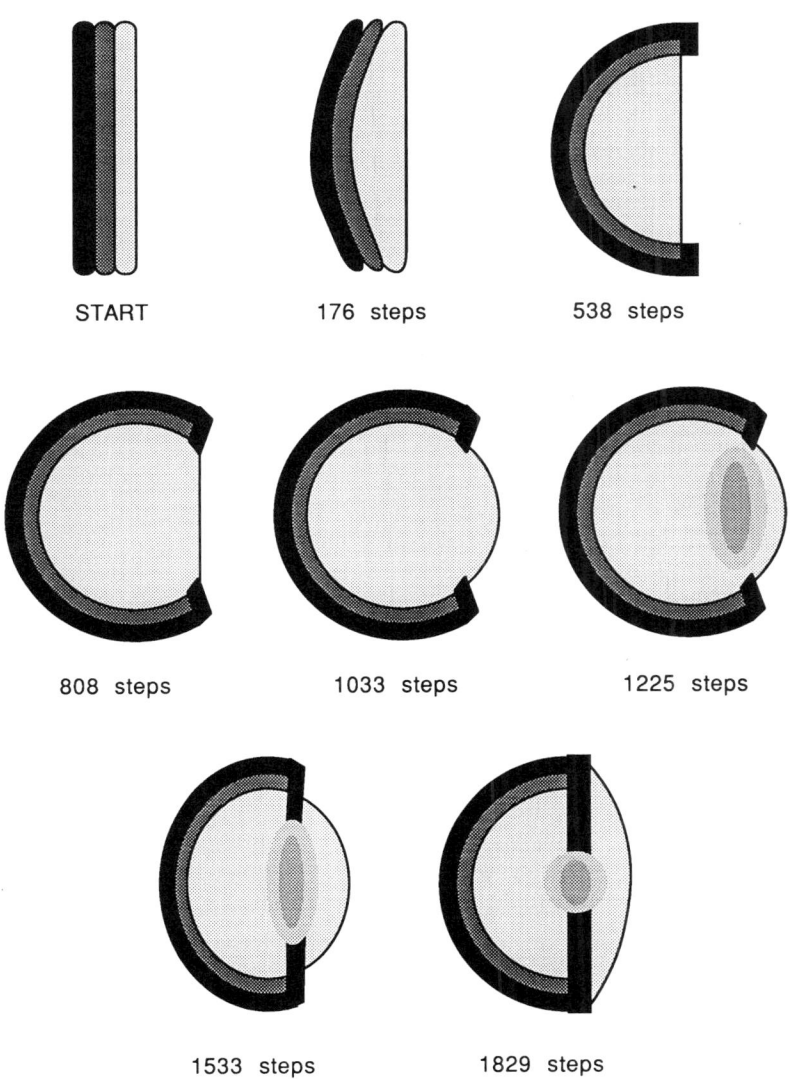

Figure 1 *Computer model of the evolution of an eye. Each step in the computation corresponds to about 200 years of biological evolution.*

once, after which they are all back in exactly the same relative positions as before. This is called a 4:2:1 resonance.

The dynamics of the Solar System is full of resonances. The Moon's rotational period is (subject to small wobbles caused by perturbations from other bodies) the same as its period of revolution around the Earth – a 1:1 resonance of its orbital and its rotational period. Therefore, we always see the same face of the Moon from the Earth, never its "far side." Mercury rotates once every 58.65 days and revolves around the Sun every 87.97 days. Now, $2 \times 87.97 = 175.94$, and $3 \times 58.65 = 175.95$, so Mercury's rotational and orbital periods are in a 2:3 resonance. (In fact, for a long time they were thought to be in 1:1 resonance, both being roughly 88 days, because of the difficulty of observing a planet as close to the Sun as Mercury is. This gave rise to the belief that one side of Mercury is incredibly hot and the other incredibly cold, which turns out not to be true. A resonance, however, there is – and a more interesting one than mere equality.) In between Mars and Jupiter is the asteroid belt, a broad zone containing thousands of tiny bodies. They are not uniformly distributed. At certain distances from the Sun we find asteroid "beltlets"; at other distances we find hardly any. The explanation – in both cases – is resonance with Jupiter. The Hilda group of asteroids, one of the beltlets, is in 2:3 resonance with Jupiter. That is, it is at just the right distance so that all of the Hilda asteroids circle the Sun three times for every two revolutions of Jupiter. The most noticeable gaps are 2:1, 3:1, 4:1, 5:2, and 7:2 resonances. You may be worried that resonances are being used to explain both clumps and gaps. The reason is that each resonance has its own idiosyncratic dynamics; some cause clustering, others do the opposite. It all depends on the precise numbers.

Another function of mathematics is prediction. By understanding the motion of heavenly bodies, astronomers could predict lunar and solar eclipses and the return of comets. They knew where to point their telescopes to find asteroids that had passed behind the Sun, out of observational contact. Because the tides are controlled mainly by the position of the Sun and Moon relative to the Earth, they could predict tides many years ahead. (The chief complicating factor in making such predictions is not astronomy: it is the shape of the continents and the profile of the ocean depths, which can delay or advance a high tide. However, these stay pretty much the same from one century to the next, so that once their effects have been understood it is a routine task to compensate for them.) In contrast, it is much harder to predict the weather. We know just as much about the mathematics of weather as we do about the mathematics of tides, but weather has an inherent unpredictability. Despite this, meteorologists can make effective short-term predictions of weather patterns – say, three or four days in advance. The unpredictability of the weather, however, has nothing at all to do with randomness.

The role of mathematics goes beyond mere prediction. Once you understand how a system works, you don't have to remain a passive observer. You can attempt to control the system, to make it do what you want. It pays not to be too ambitious: weather control, for example, is in its infancy – we can't make rain with any great success, even when there are rainclouds about. Examples of control systems range from the thermostat on a boiler, which keeps it at a fixed temperature, to the medieval practice of coppicing woodland. Without a sophisticated mathematical control system, the space shuttle would fly like the brick it is, for no human pilot can respond quickly enough to correct its inherent instabilities. The use of electronic pacemakers to help people with heart disease is another example of control.

These examples bring us to the most down-to-earth aspect of mathematics: its practical applications – how mathematics earns its keep. Our world rests on mathematical foundations, and mathematics is unavoidably embedded in our global culture. The only reason we don't always realize just how strongly our lives are affected by mathematics is that, for sensible reasons, it is kept as far as possible behind the scenes. When you go to the travel agent and book a vacation, you don't need to understand the intricate mathematical and physical theories that make it possible to design computers and telephone lines, the optimization routines that schedule as many flights as possible around any particular airport, or the signal-processing methods used to provide accurate radar images for the pilots. When you watch a television program, you don't need to understand the three-dimensional geometry used to produce special effects on the screen, the coding methods used to transmit TV signals by satellite, the mathematical methods used to solve the equations for the orbital motion of the satellite, the thousands of different applications of mathematics during every step of the manufacture of every component of the spacecraft that launched the satellite into position. When a farmer plants a new strain of potatoes, he does not need to know the statistical theories of genetics that identified which genes made that particular type of plant resistant to disease.

But somebody had to understand all these things in the past, otherwise airliners, television, spacecraft, and disease-resistant potatoes wouldn't have been invented. And somebody has to understand all these things now, too, otherwise they won't continue to function. And somebody has to be inventing new mathematics in the future, able to solve problems that either have not arisen before or have hitherto proved intractable, otherwise our society will fall apart when change requires solutions to new problems or new solutions to old problems. If mathematics, including everything that rests on it, were somehow suddenly to be withdrawn from our world, human society would collapse in an instant. And if mathematics were to be frozen, so that it never went a single step farther, our civilization would start to go backward.

We should not expect new mathematics to given an immediate dollars-and-cents payoff. The transfer of a mathematical idea into something that can be made in a factory or used in a home generally takes time. Lots of time: a century is not unusual. In fact, seventeenth-century interest in the vibrations of a violin string led, three hundred years later, to the discovery of radio waves and the invention of radio, radar, and television. It might have been done quicker, but not *that* much quicker. If you think – as many people in our increasingly managerial culture do – that the process of scientific discovery can be speeded up by focusing on the application as a goal and ignoring "curiosity-driven" research, then you are wrong. In fact that very phrase, "curiosity-driven research," was introduced fairly recently by unimaginative bureaucrats as a deliberate put-down. Their desire for tidy projects offering guaranteed short-term profit is much too simple-minded, because goal-oriented research can deliver only predictable results. You have to be able to see the goal in order to aim at it. But anything you can see, your competitors can see, too. The pursuance of safe research will impoverish us all. The really important breakthroughs are always unpredictable. It is their very unpredictability that makes them important: they change our world in ways we didn't see coming.

Moreover, goal-oriented research often runs up against a brick wall, and not only in mathematics. For example, it took approximately eighty years of intense engineering effort to develop the photocopying machine after the basic principle of xerography had been discovered by scientists. The first fax machine was invented over a century ago, but it didn't work fast enough or reliably enough. The principle of holography (three-dimensional pictures, see your credit card) was discovered over a century ago, but nobody then knew how to produce the necessary beam of coherent light – light with all its waves in step. This kind of delay is not at all unusual in industry, let alone in more intellectual areas of research, and the impasse is usually broken only when an unexpected new idea arrives on the scene.

There is nothing wrong with goal-oriented research as a way of achieving specific feasible goals. But the dreamers and the mavericks must be allowed some free rein, too. Our world is not static: new problems constantly arise, and old answers often stop working. Like Lewis Carroll's Red Queen, we must run very fast in order to stand still.

OUT OF THIS WORLD

PHOTOGRAPHS BY KENNY BEAN

Since 1994, photographer Kenny Bean has been running Photo Macro, unique photographic workshops for children and adults. He regularly takes part in the Edinburgh International Science Festival's Schools Programme, and in the Science Festival itself. Here, he talks about his photographs, his work and his plans for the future.

Elm seed

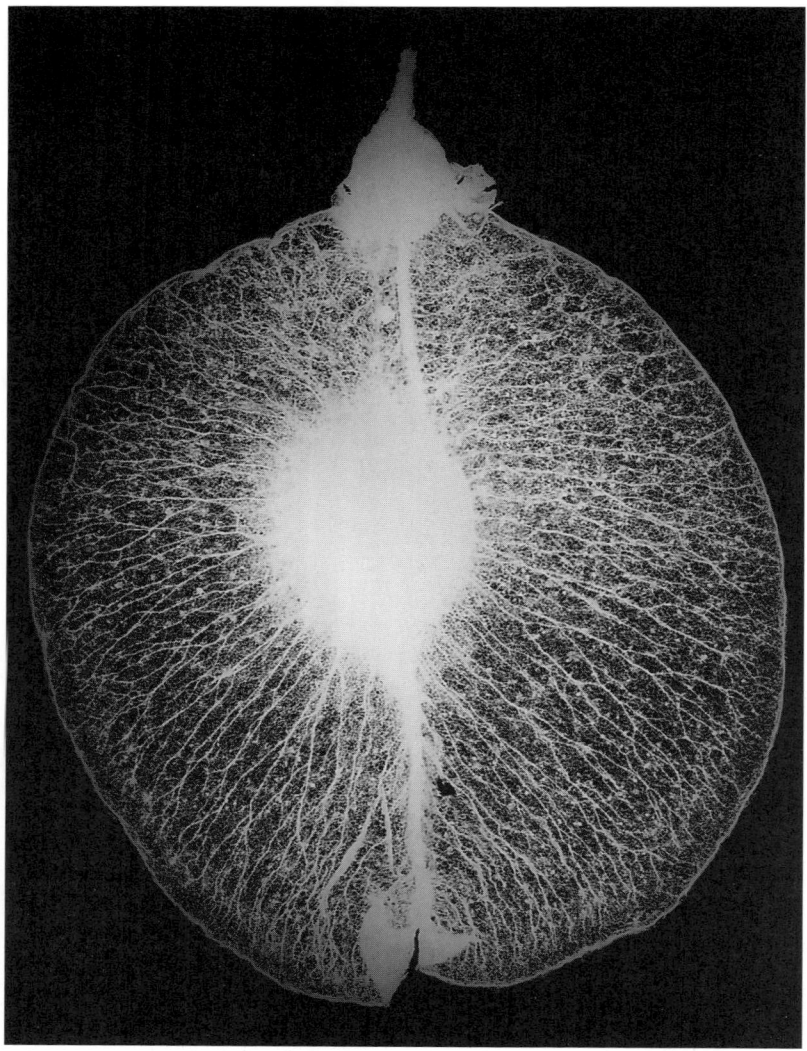

Photo Macro *developed out of demand for a workshop for people who had limited or no experience of photography. While experimenting with the use of natural objects for a children's workshop, I shone light through a feather instead of a negative in the photographic process. This gave me a high quality image. I could now get an immediate result without the need for any cameras.*

Out of this world

Feather

By taking an enlarger which is normally used to make photographic blow-ups and magnifying the size of the feather I could easily pick out the abstract patterns inside.

Feather close up

When leading the Photo Macro *workshops I see many children and adults becoming very involved and enthusiastic. When they realise how easy it is to make a perfect photograph they soon gain the confidence to explore and experiment with the materials. It is a very creative experience.*

Out of this world

Snake skin

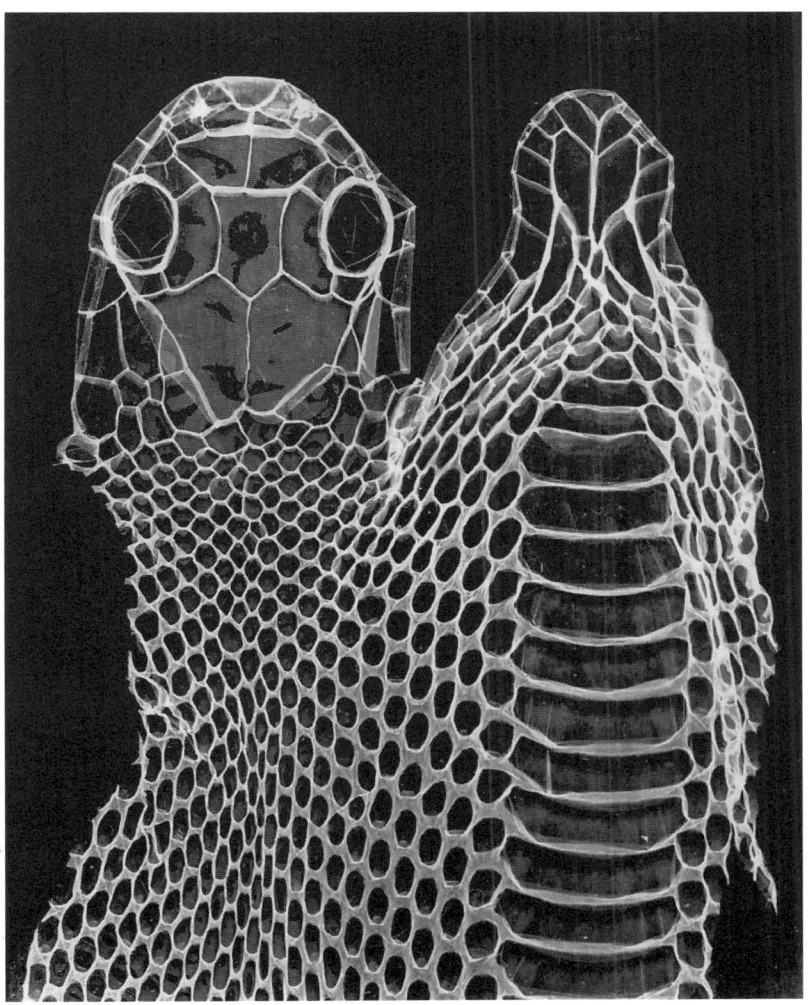

Photography without a camera makes it approachable for everyone. Many people are intimidated by cameras and technical processes. Yet there is no need to be, it is basic physics and basic chemistry. The workshop dispels the myth surrounding the photographic darkroom and introduces you to the black and white printing process.

Rhododendron petal close up

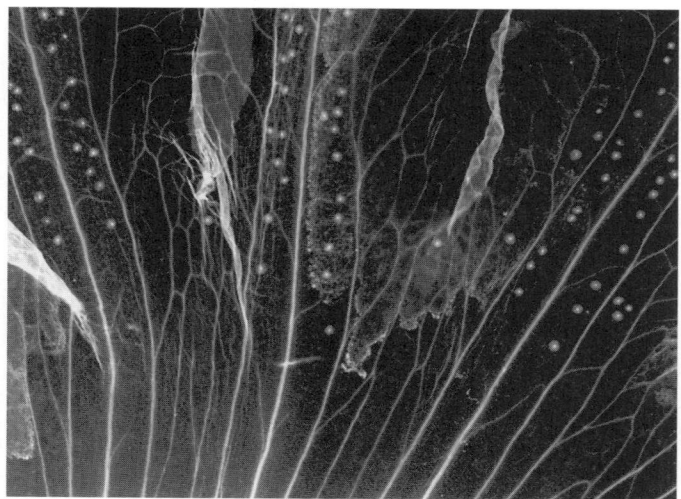

Once the basic black and white printing process is demystified, photography opens up as a creative medium and many people go away from the workshop inspired to try out their own ideas.

Rhododendron petal

The enlarger equipment I use is all from the 1960's which can be taken apart easily to show the light bulb and lens inside. These enlargers can be bought cheaply from car boot sales making it easier for anyone wanting to take up experimental photography.

Out of this world

Snake skin close up

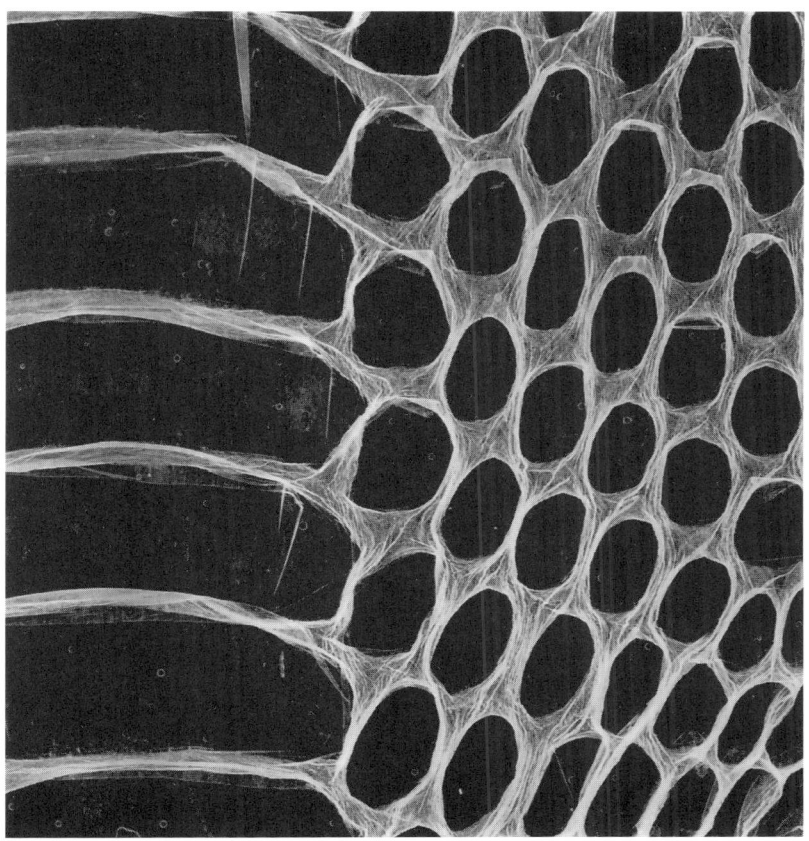

In the natural world complex patterns can be found inside apparently simple objects such as leaves and seeds. Discovering this gave me a new perspective on my surroundings. When out walking I began to look around for and explore these small objects. I found myself searching for these natural patterns everywhere.

Clematis stem cross section

The idea is to show people the patterns of growth inside these objects, to make them look closer at and appreciate the natural world under their noses. I hope it will encourage them to take a greater interest in ecological issues.

Leaf 'city scape'

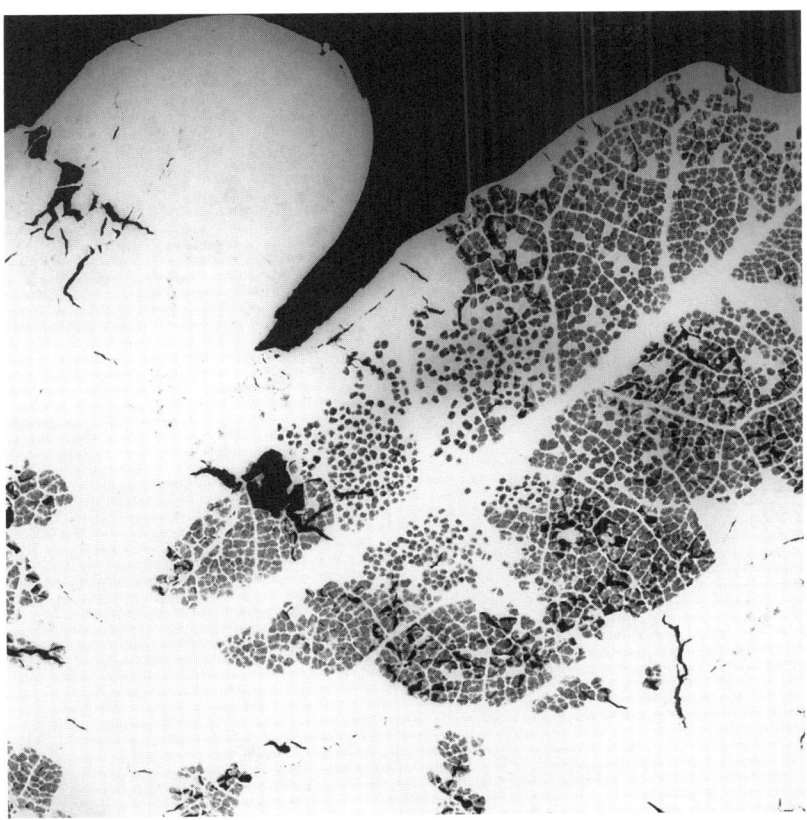

In my own work I am in the process of taking this idea further and am working on creating a series of eight foot photographs of these objects. This will allow people to see detailed abstract patterns as well as the whole object. I love the idea of someone discovering an eight foot feather.

Agate cross section

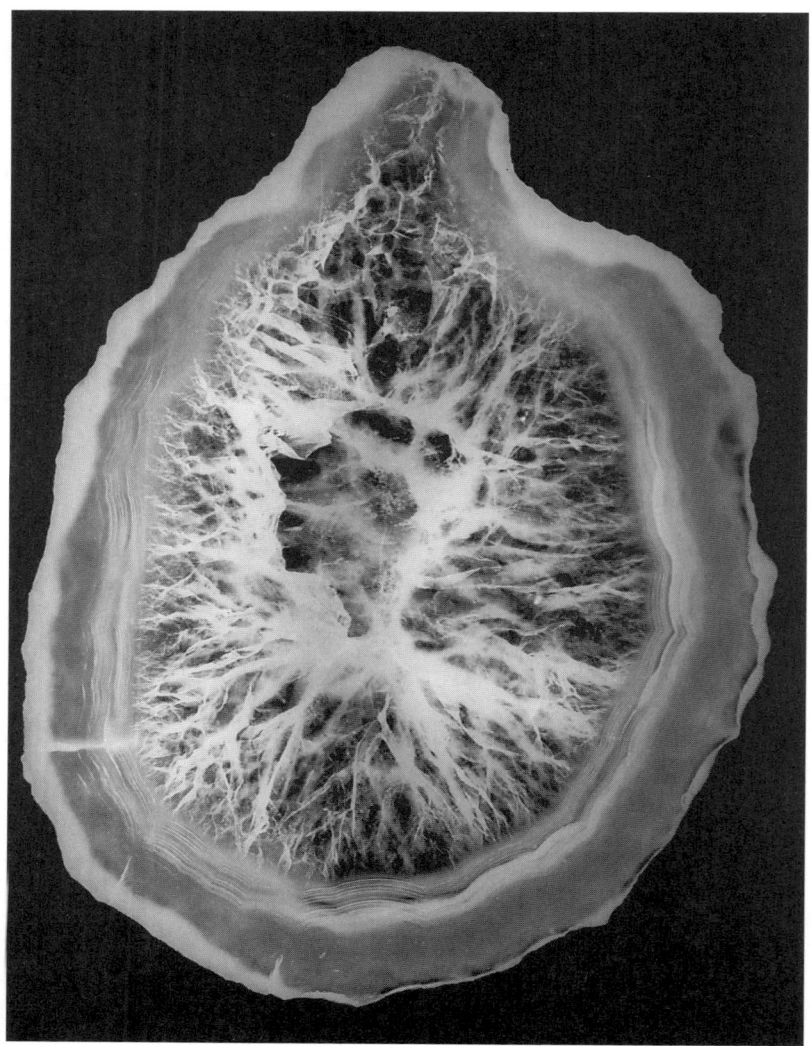

I have a theory that objects at a microscopic level relate to ariel and space images. I want to study photographs and images of the earth taken from space and compare them to close up images of objects like rocks and agates, to see if there are any similarities or connections between the images.

After working as a radio broadcaster and freelance journalist, Myc Riggulsford entered public relations with the Automobile Association before joining the health service as head of public relations for the United Kingdom Transplant Service in 1984, where he started National Transplant Week, a collaboration of 25 medical charities and organisations. Myc left the NHS in 1991 and established an independent issue management and public relations consultancy, The Walnut Bureau, specialising in medical, ethical, legal and environmental issues. It runs the Research for Health Charities Group project of the Association of Medical Research Charities, which represents 88 medical research charities spending some £340m annually on medical research in the UK. The RHCG manages issues concerning the use of animals in medical research in the press, on TV and radio, and through schools education. The group also shares information about welfare of laboratory animals. Myc Riggulsford is a member of the Institute of Journalists and the Institute of Public Relations, currently completing an MSc in Public Relations at Stirling University.

ALTERNATIVES AND THE ETHICS OF USING ANIMALS IN RESEARCH

MYC RIGGULSFORD

The controversy surrounding the use of animals in medical research is long running, very emotive, and will not be easy to resolve. On the one hand, doctors and researchers (and medical history), say that animal research has been vital in the development of some life saving treatments and improving our understanding of diseases of both humans and animals.

On the other hand, some members of the animal protection movement claim that these advances could have been made without the use of animals and that humans have no moral right to cause suffering in animals. All say that more should be done to reduce the number of animals in research, replace them with other non-animal techniques, and refine the experiments to reduce suffering and improve the welfare of the animals involved.

The year 1996 was an auspicious time to debate these different viewpoints in public at the Edinburgh Science Festival, since it was also the tenth anniversary of the law governing the use of animals in scientific experiments in the UK, the Animals (Scientific Procedures) Act 1986. It was also a year when the news was dominated by 'green' issues: veal-calf export protests, environmental concerns over oil installation dumping in the sea (Brent Spa), building new roads such as the Newbury bypass, food scares over BSE and genetically engineered tomatoes, and two major reports on the possible use of pigs as donors of transplant organs for human patients.

The use of animals in medical and veterinary research raises some key issues which are relevant to the overall way we use animals in our society, and the protection we give them; and some which are special to research uses.

According to Les Ward, Director of Advocates for Animals, without the Animals (Scientific Procedures) Act 1986, some if not all scientists engaged in animal research would "at one time or another have appeared in a court of law on a charge of cruelty to animals. And no wonder, for in the name of progress and in the search for medical knowledge and advancement, scientists have drowned, shot, burned, starved, bled, socially deprived, blinded, rendered deaf and denied sleep to, damaged the brains and severed the limbs of animals".

These are strong words indeed, but no-one taking part in the debate could deny the seriousness of the accusations when 2.8 million animals are used in research every year. The only justification for the undoubted suffering of some animals in research is that it is not cruelty caused for its own sake. None of the speakers was suggesting that researchers are cruel, causing pain or suffering deliberately and pointlessly.

In medical research which uses animals the law (and our charities which fund the research) requires that the work cannot be done any other way, that every possible effort is made to minimise suffering, and that the results of the research are important enough to justify this use of animals. Even then, it is a high price to pay. Perhaps too high. Medical research is undertaken on behalf of society, and as a society we do put the lives of humans above those of animals. When it comes to a choice, nearly all of us would be prepared to sacrifice a rat to save a person. But is the work important enough to justify this cost?

It comes as a surprise to many people to learn that the majority of the animals used in the UK in scientific procedures covered by the law every year are being used for medical research or related purposes. Many people seem to think that over half of the animals are used for cosmetic research. This is not true – only about one tenth of one percent are.

But some 90% are used in medical research: fundamental studies to understand disease; the development of new medicines; the breeding of animals as models; the production of antibodies and other products for use in research, diagnosis and treatment; safety testing so we do not expose people to harmful substances and so that we know how to treat someone who has eaten, drunk or come into contact with a dangerous product or medicine; and in surgical research and diagnosis.

Even though the critics of animal research are prepared to admit that some of the work is for important reasons, and scientifically necessary if we are to develop better treatments for humans and animal diseases, it is certainly open to dispute whether every effort is made to minimise suffering, to meet animals' complex needs, and to design experiments that reduce their use. Even if it is ethically right, do we really still need to use animals at all in medical and veterinary research, or are there now perfectly good alternatives which could be used in almost all cases?

This leads us to consider the term 'Alternatives', and look at their limitations. Alternatives refer to the various techniques which are used in medical research in addition to animal work and with which it is hoped that some animal tests will be replaced. It is unlikely that these 'Alternatives' will fully replace animal use although they may be able to reduce some animal use. At present the proportion of research which directly uses animals is quite small, although most major programmes of work will have an element of animal research amongst the other techniques used, and all the major medical research charities say animal work is necessary as an integral part of the overall research.

There are well known practical examples of experiments which used to need animals in the past but which now use an 'Alternative' method.

So if the amount of animal work is so small, and in some cases they have been replaced already, why is it not possible to do that for all experiments?

Everyone who took part in the debate, representatives of research charities and of the animal welfare groups, clearly cares about both animals and humans, especially when they are suffering from illnesses, or are in distress. Experts from every viewpoint also accept that the alternative techniques to replace all animal experiments are not yet developed.

We are not using these animals in medical research in isolation from the rest of society, and it is not just our doctors and scientists who say that it is acceptable to use animals in this way. Society has decided to exploit animals – for food, clothing, entertainment, in education (which includes medical and veterinary research), for work, and that some species of animals can be killed just for being that species – for example cockroaches, rats and mice. It is not doctors who have decided this, not our charities which pay for their work, but members of society.

It is in this context, of society's use of animals (and protection through laws governing their treatment and punishment for those who break the law) that medical research is carried out. In a society which eats animals, wears them, uses them for entertainment and work, and kills some just for being a particular species (due to the way they intrude upon our environment and compete with humans' interests), according to the medical research charities it would be morally wrong not to study them in research to save animals' and humans' lives.

The majority of animals studied in research are rats and mice – exactly the species that are exterminated for invading people's bedrooms at home. These species make up some 80–85% of animals studied in medical research. No study of disease exclusively looks at live animals, not even for veterinary diseases, but when animals do have to be used as models for humans, it is these species, rats and mice, which are studied in preference to other species such as cats, dogs, horses and non-human primates, whenever possible.

Society says different species have different intrinsic values, and that rats are generally less valuable than dogs. In veterinary research, if we were trying to find a treatment for dogs, should we only use dogs, because they will benefit, or if the same research could be done using 100 rats instead of 100 dogs, should we use the rats because they are somehow less valuable? Most people say use rats, which is what the law says.

Whenever another technique can be used instead of animal studies, it is. This humanitarian concern is supported by the law in the UK which makes it illegal to use animals if another method would do. In addition the law recognises our special relationship with cats, dogs, horses and non-human primates, and these animals cannot be used if other species would be equally suitable.

The 1986 Act also requires every project to be licensed by independent inspectors responsible only to the Home Secretary to make sure that animal use is really necessary, that the applicant is a suitable person to carry out the work, and that the welfare conditions in the laboratories are adhered to.

The term 'Alternatives' to describe non animal-using techniques of medical and veterinary research has been much misused and much misunderstood in recent years. Most commonly the research methods described as alternatives are tissue culture, computer modelling, population and patient studies, as though each, or a combination of these techniques, could wholly replace animal use, and as though our researchers unduly relied upon animal work.

In practice, this simply is not true. All the techniques contribute to scientific understanding and all research techniques have their limitations: cell and tissue culture is useful for studying particular types of cells or biological processes, but it does not tell us what happens in a complete body system. A chemical may affect cancer cells in culture, but what effect will it have on a patient's heart, lungs, liver or kidneys? For this, we need to study a living system. Rats and mice have the same organs, performing the same functions as humans' do. We may need to study effects over time, like the rejection of a heart transplant, or look at effects over several generations. A living system is recognised by our researchers as the only suitable method, and sometimes the living system chosen cannot be a human being for ethical reasons or because of the harm caused.

Tissue culture, the study of isolated cells on a laboratory benchtop, is clearly not an alternative to clinical studies of patients. In the same way research aimed at identifying differences between large populations is not an alternative to computer modelling, and it is not an alternative to the study of living animal systems either.

It is true that some of the work funded by our medical research charities has managed to replace studies which previously needed live animals with other techniques. But that does not mean that alternatives exist for all animal studies. In practice, the major medical research charities making up the Research for Health Charities Group of the Association of Medical Research Charities, which spends some £340m annually in the UK (much more than is spent on behalf of taxpayers through the government funding of research), put about 2–3% of the money directly into animal research.

This money covers the cost of animals themselves, the laboratories, housing, food and welfare, and the cost of veterinary surgeons and technicians to care for them. Although this direct cost of animals is very small compared with the money spent on other research techniques, around 20–40% of all major projects will need some element of animal use.

In most cases the newly developed techniques which have replaced, reduced or refined animal use in existing experimental methods have been developed through the continuing research undertaken by our charity funded doctors and scientists. Although money is not spent by medical research charities on specifically searching for alternatives to animal use, many of the techniques have been developed during normal research. This position was recognised and criticised as unsatisfactory by Dr Maggy Jennings of the RSPCA.

The RSPCA's ultimate goal is to see animal experiments replaced with humane alternatives – ones which cause no suffering at all, not even that caused through the distress of confinement in laboratory cages. The RSPCA is concerned about all animals, whether they are rats, mice, dogs or primates. Dr Jennings argues that medical charities are ideally placed to influence animal welfare within the scientific community, and should impose ethical and welfare criteria on all project grants. In particular she believes the need for using animals and the necessity for each individual experiment should be much more strictly reviewed. The RSPCA strongly advocates the setting up of ethical review processes at all research establishments to address these issues. But the responsibility taken by animal users should go even further, and a percentage of all research money should be set aside to specifically fund research into 'Alternatives' in order to speed up the process of animal replacement.

The Charity Commissioners, the legal authorities who govern charity actions, might not at present consider it an acceptable use of money donated for medical research to establish this sort of 'set aside' fund, since there are already animal welfare charities whose express purpose is to search for 'Alternatives'. So perhaps it is instead the responsibility of the RSPCA and other animal protection organisations to pay for any development work on alternatives. Whoever it is who should be paying, at least everyone taking part in the debate was able to agree that more funds need to be found from somewhere if the search for replacements to the use of animals is to progress more rapidly.

Meanwhile animal experiments will continue; and it is perhaps in the area of improved animal welfare that the greatest advances can realistically be made in the immediate future. It is true that the law sets minimum standards, but there is rarely a situation which cannot be improved if the real ethical dilemmas are faced up to, according to the RSPCA.

In order to do this, it is important to ask how much animals suffer, how suffering can be reduced, and whether there are ways of reducing the numbers of animals used through better experimental design.

Veterinarian Professor David Morton has some convincing examples of the improvements which can be made in animal welfare, in recognition of the fact that laboratory animals spend the vast majority of their lives not being experimented upon. For example, he advocates housing social animals such as rabbits in colonies rather than individual cages wherever possible, since studies show that in the wild they choose to spend 79% of their time with other animals. We should name them to increase our perception of their existence as individual beings with needs; vary their diets with a mixture of flavours and textures; and stimulate natural behaviours such as foraging by 'hiding' some of their food rather than providing it all on a plate.

David Morton says that 'fly-on-the-wall' natural world television programmes have enormously increased our understanding of the rational basis behind animal behaviour; and now that we have some of this knowledge, we should put our ethical considerations into action by recognising these animals' needs. If we truly believe that ultimately they will be responsible for saving many peoples' lives, then we have a moral debt to the animals used in medical research, and in order not to inflict avoidable unnecessary suffering we should do all we can to improve their welfare. For example, we should be determining and encouraging best practice, devising pre-lethal humane endpoints, and publishing information on refinement in scientific articles and journals. This information could include details of failed experiments and problems arising in the care of animals, which would assist others in avoiding the same mistakes.

Also it has been suggested that as an 'Alternative' the medical research charities should concentrate more on education campaigns aimed at preventing diseases – since it is now 'well known' that most heart disease or cancers are caused by lifestyle, lack of exercise, bad diet, and contact with dangerous chemicals. This proposal is hotly contested by the charities, on good grounds.

The research charities point out that very often it was their researchers who carried out the work to prove those links to environmental factors in the first place. Secondly, it is not as simple as that – children suffering from cancer, children with polio, ones who need heart transplants, those with arthritis, or boys with the genetically inherited disease muscular dystrophy cannot possibly be blamed for getting ill. It is not their fault. It was not due to bad diet or lifestyle. And it would certainly be wrong to punish the children and adults who become ill. So there is a clear moral obligation to continue medical research.

Thirdly, even if we all agree that more education does need to be done, it is not the remit of medical research charities to do this – if you donate money, it must be used to carry out research into diseases, not for other purposes such as education, or indeed, specifically searching for alternatives to using animals in research.

It is the medical research charities' job to try to understand diseases through research, and to carry out that work in the best way possible. If this involves improving an experimental technique, or finding a way of carrying out an experiment using tissues instead of animals, that is good, but it is not our job to search for alternatives instead of doing research.

It is techniques which lead to reductions in numbers of animals used, replacing animals with some other method and improving the welfare conditions of animals or reducing the severity of experiments which we consider to be the most important. No one technique will replace all animal use, just as no one experiment will miraculously find a cure for all cancers, or all genetic diseases. Research is a long slow process.

As practical examples of true alternatives and welfare improvements in recent years however, Research for Health Charities Group members have funded work which allows mice used in cancer research to live in group cages which better mimic their normal conditions. In Edinburgh, a test was developed for screening potential cancer causing compounds which uses fruit flies instead of mice. In Oxford and Leeds we are funding two computing projects which are generating images of heart rhythm abnormalities through modelling large groups of moving cells. But all these improvements or 'Alternatives' are developed as part of the normal research process which aims to understand disease in both people and animals, and apart from the animal husbandry improvements, they are not developed as an end in themselves.

The use of animals from a utilitarian perspective can be justified by history according to the medical research charities. It has worked in the past, for example in the development of insulin for diabetics. Every major hospital in every major town used to have a ward for polio patients; since the polio vaccine (developed with the help of animal research) became widely available, these are no longer needed. This view is disputed by members of the animal protection movement who point to examples where animal work has not succeeded, for example in failing to identify side effects of medicines. But history justifies much of the past use of animals in the view of the medical research charities. For now and for the foreseeable future we believe that we are going to continue to need to use a limited amount of animals in medical research.

However we are all very conscious of the ethical problems with animal studies, and the need to balance these with responsibilities to our public supporters and the sufferers of disease. As far as we are concerned animals should only be used when the work is sufficiently important to justify such studies; when it has been shown that animals are really necessary and the work cannot be done some other way; and when on balance, the cost in suffering and animal lives is outweighed by the likely benefit of doing this particular experiment.

To give the last word to Les Ward of Advocates for Animals, in recognition of the important part he played in the 1996 Edinburgh Science Festival debate in improving public understanding of the ethical issue surrounding the use of animals in medical research, and the limitations and possibilities of finding alternatives to their use: he quoted an American paper from Tufts University school of veterinary science saying;

"The current debate over the use of animals in research may be intense but it is largely unproductive, the assumptions that both sets of protagonists have about each other are generally false and obstruct constructive discussion. While there are always likely to be intense feelings about animal research, it is not necessary to assume that progress towards a broad public consensus is impossible. Some progress has already occurred although more by accident than design. Formal mechanisms should be established where free and open discussion of the issues that concern both sides is initiated and encouraged between both sets of protagonists".

IMPURE TRUTH

Truth
burns two-edged like a sword,
forces us into the world
from our Edens of pure research.

Each experiment
picks another forbidden fruit
turns to share it
tasting of more lost innocence.

Good and Evil
grow on the same branch
of the Tree of Life
regardless of pure Science.

Eden is over –
We have only the good earth
and each one keeper
of his brother.

Tessa Ransford

GENES, CANCER AND PREVENTION

IAN KUNKLER

Dr Ian Kunkler is Consultant Clinical Oncologist and Honorary Senior Lecturer at the Western General Hospital, University of Edinburgh. He qualified in medicine at Cambridge and St Bartholomew's Hospital, London in 1978. He trained in general medicine in Nottingham and in oncology in Edinburgh and Paris. He was appointed Consultant in Sheffield in 1988, returning to Edinburgh in 1992. He specialises in the treatment of breast cancer.

Introduction

The reduction in the mortality from cancer remains one of the major challenges for medicine as we approach the millennium. Each week fellow cancer specialists are referred patients newly diagnosed with common cancer such as those of the lung, bowel and breast. Sadly in a proportion of cases the cancer is too far advanced for cure to be possible. If we could identify and treat these cancers at an earlier stage or better still, prevent their development, the mortality of these diseases could be substantially reduced.

Ten years ago the discipline of clinical genetics had little impact upon the management of common cancers. It was largely concerned with the risk of transmission of rare inherited disorders such as Huntingdon's Chorea. The latter is characterised by a disorder of movement and progressing to dementia and early death. What has happened in the last 10 years is little short of a revolution in the understanding of the genetic basis of cancer. It is now appreciated that most cancers in man arise from a series of genetic alterations in a single cell. As cells divide they are at constant risk of genetic mutation. It is these mutations which may disturb the regulatory processes which control normal cells. Accumulated genetic damage may result in the daughter cells being unresponsive to the signals from surrounding cells and replicating in an uncontrolled fashion, one of the cardinal features of a cancer or malignant tumour.

Landmarks in cancer genetics

The idea that cancer might have a genetic basis is not new. In 1912 Bovert, a German cytologist, observed that malignant cells had abnormal chromosomes. He thought that any factor leading to this abnormal appearance could cause cancer. With improved staining techniques to identify individual chromosomes in the 1970s it was appreciated that chromosomes from cancers were often broken with parts of one chromosome joined to others. In some cases, whole chromosomes were absent. However, until the 1980s it was uncertain if these chromosomal changes were the primary events that led to cancer or they were secondary to the development of cancer.

It was the development of specific scientific techniques for resolving this uncertainty that were a major landmark in establishing the genetic basis of many cancers. Over the last two decades many of the genes that are involved in the transformation from a normal to a malignant cell have been identified. The key developments have come from molecular biology. New methods allow us to study the DNA (deoxyribose nucleic acid, the cell's genetic material) at a molecular level and how it functions. This new research has provided insights into familial predisposition of common cancers such as those of the bowel and the breast. In addition, it has provided clues to the individual variations which we see in the clinic in the response of normal tissues such as the skin to the standard doses of radiotherapy (ionising radiation) used in the treatment of cancer. This article concentrates on summarising some of the developments in cancer genetics relating to common cancers of the breast and bowel.

Human Genome Project

The key to unlocking this new knowledge have been major technological advances in the study of genes at a molecular level. A major stimulus has been the Human Genome Project. This is an international collaboration between scientists to map the whole of the human genome. This is the largest scientific endeavour in the history of science. One of the main aims of the project is to provide detailed 'maps' of particular molecular landmarks on all the 23 human chromomes. These are known as 'sequence-tagged sites'. The geneticist can use these markers to identify particular sections of DNA, the cell's nuclear material. DNA is composed of four different bases – adenine, guanine, cytosine and thymine. It is DNA that codes for the construction of particular proteins. The project is planned to take 15 years, ending in 2005. The complete set of human genes is thought to contain between 50,000 and 100,000 genes. Of these, only a few thousand have so have been identified, but the number is rapidly expanding. The aim is to clone most of the human genome by the next century. As genes are cloned, it is possible to test their candidacy in causing particular cancers. The identification of so called 'cancer susceptibility' genes is casting light on fundamental aspects of cancer causation. However, the identification of individuals who are at increased risk of cancer poses major challenges for the cancer specialist and the geneticist.

The genetics of cancer

Genetic predisposition to common cancer is relatively uncommon. For example, it only represents about 5% of breast cancers. Most cancers arise spontaneously (so called sporadic cases). Clinicians have long been aware of some patients with breast cancer having a family history of affected relatives, usually mother, sister or aunt. The analysis of breast cancer at a molecular level has revealed many abnormalities. These may be steps along the pathway towards malignant transformation. These genetic abnormalities include loss or gain of whole chromosomes. There is much evidence to support the idea that genetic abnormalities are acquired sequentially over time. The best evidence for this is from patients who have the genetic abnormality in which the patient develops many polyps in the lining of the bowel (adenomatous polyposis coli). In time, some of these benign polyps may turn malignant, often many years later. Carcinogenesis, the process by which cells become malignant, may result from alterations in a relatively limited number of biochemical pathways in the cell. If we understood how these normal pathways had been altered, it might be possible to repair them. In so doing progression from benign disease to frank cancer might be prevented.

It seems paradoxical that if a genetic defect is passed on through the male sperm or the female egg it must affect all the cells of the body. However, in cancer-prone families cancers were observed in a restricted number of tissues and sometimes only one tissue. This implied that more that one genetic 'hit' was needed. Knudson working in Philadelphia first proposed the 'two hit' theory. Essentially this meant that the first genetic defect predisposes to the development of malignancy. However, it does not actually cause the cancer. A second genetic abnormality ('hit') is required to give rise to cancer.

Oncogenes and the development of cancer

Many of the changes that occur in malignant cells are due to mutations in the DNA sequence. Mutations are alterations in DNA due either to substitution of one pair of bases (point mutation) or translocations where there is an exchange of genetic material (chromatin) between chromosomes. Certain gene sequences seem to give rise to malignant change in human cells. There are known as oncogenes. It had been known for many years that certain viruses can cause cancer in animals. It was subsequently shown that many of the genes suspected of involvement in human carcinogenesis were carried by viruses. It also emerged that many of the genes causing cancer in man had been acquired by viruses.

We believe that cancer occurs either because of mutations occurring in normal cellular genes or because viral oncogenes have been introduced. The normal version of these genes (so called proto-oncogenes) code for proteins which stimulate the cell to divide. It was subsequently appreciated that mutations in proto-oncogenes might transform them into carcinogenic oncogenes. These mutations cause the cell to produce abnormal amounts of certain proteins or abnormal proteins.

Tumour suppressor genes are normal cellular genes which inhibit malignant transformation. However, when mutations occur in these tumour suppressor genes, their normal function is lost. The best known example of a tumour suppressor gene is p53, the so-called "guardian" of the genome. In many human cancers mutations have occurred in p53 with loss of its function as a tumour suppressor gene.

Importance of cancer genetics on bowel and breast cancer

While genetic predisposition may only account for perhaps 5% of the total incidence of malignancy, some of the cancers which are known to have a genetic basis, albeit in a minority of patients are very common (*e.g.* bowel and breast cancer). The total number of women worldwide with breast or colorectal of genetic origin is therefore likely to be substantial. For example, the lifetime risk for breast cancer in a women is 1 in 12 and there are over 500,000 new cases a year. In the UK, there are over 30,000 new cases a year. Cancer of the breast accounts for 18% of all female cancer. The incidence is increasing by about 1–2% per year, especially in elderly women. 40% of those diagnosed will die of the disease (15,000 per year in the UK). On the basis of 6% of breast cancers being of genetic origin, 1 in 200 women with breast cancer have a genetic susceptibility. It is estimated that about 150,000 women in the UK carry a breast cancer susceptibility mutation.

Cancer of the bowel and rectum affects one in 34 people, and one in 52 people will die of the disease. There are over 280,000 new cases of cancer of the bowel and rectum each year worldwide. In the UK, there are over 14,000 deaths from cancer of the bowel. Although most cases of bowel cancer are not inherited, the gene that is mutated in the familial type of this cancer is also mutated in the commoner non inherited form of the disease. Thus all forms of bowel cancer may share some genetic features.

Clinicians have a long tradition, predating the recent advances in cancer genetics, of documenting any family history of breast and other cancers. In breast cancer, it was not uncommon for women to report cancers in their mothers, grandmothers or aunts. However we had no knowledge of the nature of the genetic links between the same cancers among close relatives. Patients often have hazy memories of the diseases afflicting family members. They may have only limited information about cancers in the family, particularly when they themselves may only have been children. Cancer was often not talked about as a taboo subject. We know how inaccurate a family history taken from a patient can be for the true incidence of cancers in a family can be checked through the Public Records of Birth, Marriages and death. Discordance is often found between the patient's account and that recorded in the Public Records. In future our Public Records system will be of great importance in identifying the pedigrees of patients with cancer predisposing genes where other family members may needed to be alerted by doctors to an increase risk of cancer.

Breast Cancer

Several breast cancer susceptibility genes have been recently been identified. The first to be identified was BRCA1 (breast cancer 1). It was found on part of chromosome 17. An international collaborative study showed that it accounts for virtually all patients who have cancers both of the breast and the ovary. Carriers of the BRAC1 have an 80% risk of developing breast or ovarian cancer by the age of 70. It is estimated that about 25,000 women in the UK carry BRCA1 mutations. In 1994, a second gene BRCA2 was identified. It lies on chromosome 13. It is thought to lead to approximately the same proportion of early onset breast cancer as BRCA1. There is some evidence that there is a third breast cancer susceptibility gene, BRCA3.

Sensitivity to ionising radiation

A proportion of breast cancers may be predisposed to by carriage of the gene for a rare disorder known as ataxia-telangiectasia (A-T). It is characterised by impaired immune function, an excess sensitivity to ionising radiation and susceptibility to malignancy. In particular female A-T heterozygotes have a 6–8 fold increase in breast cancer. It has been estimated that up to 9% of women in the United States may be heterozygotes for A-T. Recently the gene for ataxia-telangiectasia was cloned. It should therefore, be possible to establish what proportion of women with breast cancer actually are heterozygotes for A-T. Previous estimates may have been unduly high.

Ionising radiation (radiotherapy) is an increasingly used local treatment of the breast following the removal of a small breast cancer. A small dose is given each day over periods of 3–6 weeks. If, as is possible, the radiation sensitivity of the cancer to radiotherapy parallels that of the surrounding normal tissues that are also irradiated, it may be possible to treat some patients with cancer who are genetically predisposed to a more intense radiation reaction, with lower doses without compromising the chances of cure. Acute and long term side effects of radiotherapy might be also usefully be diminished. Clinical experience suggests that patients with unusually intense radiation reactions are relatively uncommon. However, the benefits even to a minority of patients could be substantial.

Genetic predisposition in bowel cancer

It has long been known that some families have several individuals with multiple polyps (excrescences from the inner lining of the wall of the bowel) which eventually become cancerous. This syndrome is called familial or adenomatous polyposis coli (APC). It accounts for a only small proportion (2%) of the patients who develop cancer of the bowel. The molecular study of bowel cancer has identified a sequence of events in the progression from benign to malignant disease of the bowel. Polyps have the advantage that they are amenable to being viewed directly by the physician or surgeon and removed for examination under the microscope. Families with familial polyposis coli could, therefore be studied in detail. There was clearly an inherited defect which first gave rise to the formation of multiple polyps and to cancer or the colon. It was demonstrated by Bodmer in London and Leppert in Utah that the APC gene lay near the middle of chromosome 5. Its precise position was subsequently identified by Vogelstein at Johns Hopkins University and by Nakamura in Tokyo. It was shown that all the cancer related mutations in the APC gene gave rise to an incomplete protein. Vogelstein also demonstrated that cancers of the colon invariably include the genetic mutation found in the polyp. However, the cancer contains additional mutations, accounting for the malignant behaviour of the cancer.

The importance of a cancer to public health

From a public health point of view, cancer represents a major cause of death. Most people will know of someone who has died of cancer. Over 30,000 new cases of breast cancer in women in the UK are registered each year. Of these 40% will die of the disease. Breast cancer accounts for 10% of deaths in the western world. To date no environmental factor has been clearly identified as a cause of breast cancer.

Potential of genetics for early diagnosis

The outcome of treatment is generally more favourable when these cancers are diagnosed at an early stage. Tumours may grow without giving rise to symptoms. Sometimes the first indication to the patient that there is something seriously wrong may be when the cancer has spread to other organs such as the liver causing loss of appetite. Cancers of the bowel limited to superficial layer of the bowel wall have a very high cure rate (over 80%) with surgery alone. Early cancers confined to the breast of less than one centimetre in diameter have a very good outlook when treated by surgical excision and radiotherapy with recurrence rates of 5% or less at 5 years after treatment. Breast screening to detect breast cancer early, for example, has reduced the mortality of breast cancer in women over 50 in Scandinavia by 30%. While the proportion of breast and bowel cancer that are due to an inherited predisposition is small (*e.g.* about 10% for breast cancer), efforts to identify those are risk to facilitate early diagnosis and treatment are of great importance. If we could identify individuals at a high risk of developing common cancers, it may be possible to devise treatment which might prevent the development of these cancers. The potential benefits of such approaches, if successful, would be enormous.

The psychological effects of genetic risk
Coping with genetic risk
At one extreme individuals at high risk of developing cancer may live in lifelong fear of cancer, unable to cope and make the most of life's opportunities. For this reason, detailed information is needed on how individuals cope who have been voluntarily tested for a cancer predisposing gene (often to assess the risk of siblings or children). Answers to these questions are being sought by clinical psychologists working closely with cancer genetics clinics. How do levels of anxiety and depression correlate with differing degrees of genetic risk and with the psychological make up of the person? How effective is psychological support in helping the patient cope with an uncertain future? Research in this area is still in its infancy.

Ethical and insurance issues
The Health Insurance industry is primarily concerned with the assessment of medical risk. Cancer is generally a disease of the middle aged and elderly and very often reduces life expectancy. One of the problems of genetic risk of cancer is that one is dealing with the probability rather than the certainty of developing cancer. A patient carrying the BRCA1 gene might developed a cancer in their 40s or their 70s. Developing a cancer of the breast earlier in life has more profound financial consequences for the patient and her family through loss of earning power through ill health.

The potential for screening individuals for the presence of cancer predisposing genes raises major ethical issues. Many have an interest in the outcome of the debate on the application of genetic testing. Who should decide on these issues – the doctor, the government or society at large? At present, there are more questions than answers. Those who carry cancer predisposing genes for cancers where there is no evidence that earlier treatment necessarily is more effective or who might develop a cancer very late in life might well feel that their and their family's lives and employment prospects are severely blighted. What weight does one give to the interests of children or future generations in deciding whether an individual should be screened? Does an individual who has tested positive for a cancer predisposing gene have a moral responsibility to notify other family members who might opt for screening of their own genetic test result? Would it become an offence in law to withhold such information from interested family members?

The confirmation that the individual and their relatives may be at risk of life-threatening cancers has profound consequences for all of them. Some would argue that genetic testing should only be carried out if it affects the management of illness. Others suggest that this genetic information should not be available to insurers since it would be discriminatory, unfairly penalising those whose cancer prone inheritance was no fault of their own. At present, the insurance industry does not seem to be planning to introduce genetic testing for cancer as a screening device in assessing insurance risks. These matters are being considered by the Research Committee of the Life Insurance Board. They are examining among other issues the philosophical issues and the practical implications for the life insurance industry. Actuaries perceive fairness in terms of 'equal treatment for equal risks'. However, fairness from the actuaries perspective may differ from concepts of fairness of other groups in society.

At present in the UK, insurance plays a much smaller role in health care than in the United States. However, the funding of health care in the UK is changing with increased involvement of the insurance industry in long term care of the elderly. Perhaps the state should become more directly involved in underwriting those at increased genetic risk of cancer.

Potential benefits of genetic knowledge

The information gained from testing an individual for the presence or absence of a cancer predisposing gene is not limited to this question alone. Patterns of genetic mutations may have important consequences for the diagnosis of the disease or the outcome of treatment. For example, where genetic mutations follow a well defined sequence, it may be possible to assess (stage) the extent of the cancer. This may help in the appropriate selection of treatment. If we know the genes that are mutated, it may be possible to detect cancer cells in tissues before they grow to a size that is detectable clinically or on radiological imaging. It may be possible to design drugs that counteract the effects of particular mutations. This might result in arresting the growth of a cancer temporarily or even eradicating it. It may not be necessary to correct all the mutations since only one or two genes may be responsible for the formation of some cancers.

Conclusion

The revolution in the understanding of cancer predisposition brought about by modern molecular biology raises major ethical and economic issues. It is obvious that in the UK a state funded national network of cancer genetics clinics is needed, staffed with doctors, nurses and psychologists with the skills to provide the patient with the very best up to date advice. To date many of these cancer genetics clinics have been funded from the major cancer research charities. In many areas of the country, the service provision of cancer genetics has yet to be funded. Services are, therefore, patchy.

The advances in the cancer genetics offer hope for improving the earlier detection and outcome and potentially for preventing some cancers in individuals shown to be at risk.

Decisions upon how genetic testing for cancer predisposing genes should be applied needs widespread consultation within the National Health Service, the Law, the Life insurance industry and society as a whole. The right course is not easy to identify. However, whatever guidelines are adopted a sense of common ownership and support by the professions and the society they serve is needed.

Sir Crispin Tickell GCMG KCVO is Warden of Green College, Oxford; Chancellor of the University of Kent at Canterbury; Chairman of the Climate Institute of Washington, DC; Director of the Green College Centre for Environmental Policy and Understanding; Chairman of the Government's Advisory Committee on the Darwin Initiative; and Convenor of the Government Panel on Sustainable Development. Sir Crispin was a member of the Diplomatic Service. He was Chef de Cabinet to the President of the European Commission (1977–80), Ambassador to Mexico (1981–83), Permanent Secretary of the Overseas Development Administration (1984–87), and British Permanent Representative to the United Nations (1987–90). He was also President of the Royal Geographical Society (1990–93), and Chairman of the International Institute for Environment and Development (1990–94). He is the author of Climatic Change and World Affairs *(1977 and 1986), and* Mary Anning of Lyme Regis *(1996). He has contributed to many other books on environmental issues (including human population increase and biodiversity). His interests go very wide, and include business and charities. More personally he cherishes mountains, pre-Colombian art, and palaeo-history.*

THE ENVIRONMENT: A MARRIAGE OF DISCIPLINES

CRISPIN TICKELL

Environmental issues are like natural ecosystems: a tangle of multiple, ever changing, interacting factors of almost infinite complexity. General understanding of them involves many disciplines and combinations of discipline, ranging from scientific observation and mathematical assessment of risks and uncertainties to economics, history, sociology, politics and all except the most extreme specialisations.

It sounds easy but it is of course amazingly difficult. There are few brave enough to have an integrated or world view. For one thing no one likes to be wrong, and world views change continually, and change is always painful. So the frustration, particularly strong among scientists, is to become the master of a speciality, and linger comfortably in the half light at the bottom of a box of secure knowledge. Science festivals apply leverage to the lids of boxes and let in the daylight. Out come the occupants, blinking a little and looking at each other, not always with enthusiasm, but at least with interest.

Only an inter-, or intra-disciplinary approach can make sense of the environment. It also requires a stretching of time and space, for which our genes are perhaps ill adapted. Our minds have difficulty in grasping the long term, and in particular in bringing it into the short term. They are likewise discommoded by the sheer speed of change – the acceleration of change – which marks our times. Paradigms seem to shift at such a rate that we hardly have time to understand each shake of the kaleidoscope. But we have to try. The problems are becoming pressing. They reach into every part of the human condition.

Global environmental change

Natural change

Change is inherently natural. We ourselves are the products of the undulations of the ice age. As an animal species we did not appear as such until very recently. Most change is slow enough to permit ecosystems and the individual species within them to adapt. The last 10,000 years have been remarkably stable, but overlying natural change have come the accumulating effects of human induced change.

Human-induced change

Like many other animals, we have, usually unwittingly, changed our environment, but so far such change has been mostly local or regional. Now the situation is altering. The industrial revolution, which began around 250 years ago, has been the great accelerator. Economic wealth on the normal – highly misleading – definition has risen at an almost incredible rate. But not only has this growth been highly uneven, it has been at a price. We have created what future generations will surely find a peculiar society, hooked on fossil fuels, and pulled by a consumerist philosophy out of synchrony with the natural world. This way of life cannot be sustained, and yet it appears to be inescapable. Let us consider six of its aspects which lead into each other and cannot be dealt with in isolation.

First, the growth in human numbers: from less than 10 million 10,000 years ago, to a billion in the time of Robert Malthus, to 2 billion in 1930, and to around 5.8 billion today. The human population is now increasing by over 93 million a year (or a new China every 12 years). It will almost certainly rise to 8.5 billion by 2025.

Second, degradation of land has occurred. According to the United Nations Environment Programme's Environmental Data Report for 1993-94, some 17% of the world's soils have been damaged by human activity. Enough good topsoil is lost around the world every year to cover the agricultural regions of France. 16% of the landmass of the former Soviet Union was judged an ecological disaster area by the now defunct Soviet Academy of Sciences. The amount of dust and other aerosols in the atmosphere, some of it caused by human activity, may already be affecting climate.

A third aspect of human-induced change is pollution of seas, rivers and fossil water. Urban sewage and industrial effluents are a major threat to rivers and coastal waters and in many areas of the world the quantity has outstripped ability to cope. Pollution of fresh water is all the more serious at a time of rapidly increasing demand. Such demand doubled between 1940 and 1980, and is likely to rise again between 1980 and 2000. The amount of fresh water available has not changed since Roman times; but the human population has risen from 200 million to 5,800 million, and human needs with it. Even in the remotest seas, such as the Arctic and Antarctic, significant quantities of toxic chemicals – polychlorinated biphenyls, DDT, lead and cadmium – have been found in air, seawater, sediments, fish, mammals and seabirds.

Fourth, changes in the chemistry of the atmosphere have occurred. Acidification is local in character, and can be remedied with money and political will. Ozone depletion has global consequences. More important than the danger of increasing human melanomas is the potential effect on other organisms from vegetation and crops to phytoplankton in the ocean.

Next climate change is the least predictable, and potentially the most serious aspect of all. Global warming enhancing the natural and indispensable greenhouse effect could affect every aspect of human society. To judge from recent, still controversial evidence, it could also destabilise the system. The main conclusions of the Intergovernmental Panel on Climate Change (1990, updated 1992, 1994 and December 1995) represent a broad scientific consensus. On the assumption that we continued to pump carbon dioxide and other greenhouse gases into the air at current rates the Panel concluded there could be a rise of global mean temperature of around 2°C by the end of the next century. Compare this with a drop of around 5°C during the last glacial episodes. The results would of course be very different in different places, and Western Europe could be cooler if the oceanic circulation system were to be changed. Estimates of sea level rise are less precise but it could have risen by around half a metre by the end of the next century. But averages are misleading: effects will be much greater in some areas than others, with major disruptions of weather systems, especially in regions already at climatic risk.

Last, depletion of the diversity of life has occurred: nearly all the species that have ever lived are now extinct. But they did not die out at once. Humans, as they alter and destroy whole ecosystems, are causing extinctions at up to 1000 times the normal rate. As E.O. Wilson has written – "a fifth or more of species of plants and animals could vanish or be doomed to early extinction by the year 2020 unless better efforts are made to save them". Our understanding of the complexity of ecosystems is very limited and we have little idea which are more important to us than others.

The rise in public awareness

It is too late to prevent all these problems; but over time they can be mitigated and we can adapt ourselves to them. Over the last quarter century attitudes towards them have changed substantially. The road from the UN Conference on the Environment at Stockholm in 1972 to that on Environment and Development at Rio de Janeiro in 1992 has many landmarks: the creation of the UN Environment Programme; the Global 2000 Report; two World Climate Conferences; the Brundtland Commission Report; the Vienna Convention, Montreal Protocol and subsequent agreements on ozone; and countless conferences, seminars, workshops, books and studies on environmental issues and sustainable development. With the Rio conference of 1992 which produced the Conventions on Climate Change and Biodiversity, Agenda 21, and the UN Commission on Sustainable Development, with the reconstituted Global Environment Facility, with the UN Conference on Population and Development in 1994, with the long awaited implementation of the Law of the Sea last year, governments and people the world over now have the mechanisms with which to cope if they have the will to use them.

Change in thinking

It is all too human to resist change as long as possible, and to modify rather than reconstruct. The conventional wisdom is dead: long live the conventional wisdom. But trying to understand environmental change requires change in how we think. As anyone knows who has tried to change thinking even on one specific issue, there is an enormous weight of inertia, deriving from countless sources, in favour of remaining within the framework of current orthodoxy. Lord Keynes once said: "The difficulty lies not in new ideas but in escaping from old ones." Thus governments will sign such Conventions as those on Climate Change and Biodiversity, but continue to operate in terms inconsistent with them. I give two examples:

Economic growth

There is almost no politician nor economist in an industrial country who does not extol the virtues of economic growth. Growth will pull us out of recession, solve unemployment, put the colour back into our cheeks. But growth in this traditional sense is most misleading. It takes no account of the impact of growth on the environment nor of the accompanying loss of resources. Economic activity has often been seen as a circular flow of exchange value on an ever increasing scale. But it also entails the consumption of non-renewable resources and the production of wastes. This is rather as if biology tried to understand animals only in terms of their circulatory systems without recognising that they had digestive tracts as well.

Thus we should recast our vocabulary and say more exactly what growth should be. It surely means the optimum management of our economies within the limits of sources and sinks on the basis of three principles: the rate of renewable resources (soil, water, forests) should not exceed the rate of regeneration; the rate of use of non-renewable resources (fossil fuels, minerals, fossil groundwater) should not exceed the rate at which sustainable alternatives can be developed; the rate of emissions of pollutants should not exceed the capacity of the environment to assimilate them.

Development

My second example is the use of that other prince of cant phrases: development. Again there is almost no politician or economist, this time from non-industrial countries, who does not sing a hymn to development. UN documents are littered with references to "rights to development". Development is almost universally seen as the best, if not the only, escape from poverty. For most it means becoming as much like industrial countries as possible with broadly the same living standards. Unfortunately, with the world's assets distributed as they are, and with pressures arising from population increase and other hazards, there is not the slightest prospect that this will ever happen. Expectations far exceed reality, with dangerous consequences for those who nourish them. Again redefinition is urgently required. In my view development should be seen as the realisation of potentialities, whatever they may be in different parts of the world. The end result is not much different from redefined growth.

The changes which are needed in ways of thought were well set out in the report of the Brundtland Commission on Environment and Development in 1987. In this sustainable development was defined as "meeting the needs of the present generation without compromising the ability of future generations to meet their needs". I prefer my own definition: "durable change for the better while protecting the earth we inherit and the earth we bequeath".

Three principles need to be put into practical effect. The first is the precautionary principle or approach, well defined in the Declaration to which all governments committed themselves at Rio: "where there are threats of serious or irreversible damage, lack of full scientific certainly should not be used as a reason for postponing effective measures to prevent environmental degradation". The second is the principle that the polluter should pay (adopted by the OECD countries as long ago as 1972). Finally the third principle is that environmental considerations should be brought into the centres of decision making at all levels.

Where do we go from here?

There are three broad aspects. First we need more impartial, long-term, large-scale, interdisciplinary research.

Second, better understanding of the science must mesh with better understanding of the political, economic and social aspects. The practical implications of the changes which are taking place require far more attention than they have been given. We need not so much doomsters as a new Domesday Book to record and so far as possible look forward. People could thus see for themselves the relationships and connections between things, and link them with prospects for the future. I want to distinguish four of particular importance. In a sense they constitute pressure points. Each arises from the combination of variables of increasing human population, degradation of land and water, climate change, and depletion of biodiversity. I look at each in turn:

- First is the current measure of disruption within societies. The age of nation states was to some extent succeeded by the age of ideologies; and the age of ideologies has been succeeded – so far – by an age of anxious uncertainty. The end of the cold war was like the ending of an ice age. Once the weight of the ice had been removed, landscapes and pressures which had long been forgotten were revealed. People who had lived well enough together under the disciplines of the nation state or on each side of the Communist/capitalist divide suddenly rediscovered ancient grievances and animosities. Add in some cases the forcing factors of population increase and environmental degradation, and the result has been loss of social cohesion and respect for law, whether in the former Soviet Union and Yugoslavia, or in Rwanda, Somalia, Haiti and Nigeria. Cultures, groups and communities with their own characteristics have come to be seen as more important than national governments, themselves often in debt or other economic trouble. It is hard to see what can be done about this range of problems. The international community represented by the United Nations has so far shown itself to be unable to cope.

- Next is the problem of human migration. In the last fifteen years there has been a steep rise in human displacements. In 1978 there were less than six million refugees on a strict political definition; by 1994 the figure had risen to over 22 million. There is also an unquantified number of environmental refugees: people who can no longer gain a secure livelihood in their homelands because of drought, soil erosion, desertification, deforestation and other environmental problems. Some move across frontiers, others are displaced within them, but depending on their definition the number could be as high as another 22 million. No wonder that the refugee problem is now a daily headline, and that the barriers against them have been going up all over the world. Again with environmental forcing, the problem is likely to get worse.

- Following this is the threat to human health. We must expect changing patterns of disease. Temperature and moisture are determining factors for living organisms in soil, air and water. Variations in both affect the ability of viruses, bacteria, and insects to multiply and prosper. These organisms are able to reproduce rapidly and therefore can rapidly adapt to changing environmental conditions. One illustration of this is the development of resistance to modern drugs and pesticides. Examples include the rise in antibiotic-resistant bacteria, and drug resistant malaria.

The re-emergence of such diseases as cholera, dengue, and other viral fevers, can be traced to changes in land use, human mobility, and the growth of cities. As in Europe until the last century, infectious diseases thrive in urban populations. The frequency of contact, the density of the population and the concentration of infective and susceptible people promote the transmission of disease. There are also indirect effects. It was surely more than a coincidence that the outbreak of the Black Death in the fourteenth century had its primary impact on populations already weakened by lack of food arising from shortages from the beginning of that century. Such considerations are of particular relevance when dealing with biotechnology and, in particular, the release of genetically modified organisms into the wider environment. Here, more than in other areas, there is uncertainty about the long-term outcome of human actions and of human ability to deal with the consequences.

- The final point is the gap between rich and poor. That too is becoming wider. Whatever the problems of the industrial countries, they are as nothing to those of the rest of the world. In 1960 the richest 20% of countries enjoyed 70% of global income: by 1989 this proportion had risen to 83%. In 1960 the poorest 20% of countries enjoyed 2.3%: by 1989 this proportion had fallen to 1.4%. Somehow issues of high consumption (and with it high production of waste) in some parts of the world and increasing poverty in others have slipped from the top of the world's agenda, even at Rio. Again, with environmental forcing, the problem is likely to get worse.

Some people still comfort themselves with the thought that these are other people's problems. Social justice for others is rarely a vote winner. Nor is concern for other people's health. But even if some people and governments wished to seal themselves off from the rest of the world, they could not do so. In no continent, country or city can the rich fortify themselves for long against the poor. Land frontiers can always be penetrated: look what has happened across the long southern frontier of the United States, which has already become markedly more Hispanic as a result. Nor are sea crossings a real barrier. Desperation could push Arabs and Africans north across the Mediterranean, the Chinese into the empty spaces of Siberia, and the Indonesians into northern Australia, each bringing their micro-organisms with them. A combination of the four factors I have mentioned – loss of social cohesion, migration, disease, and poverty – could change the face of the 21st century.

My third general thought is the need to communicate this interdisciplinary understanding to a wider audience: policy makers and the public. Intelligibility is no academic question. It matters deeply to us all. It is always a matter of wonder to me how a society so profoundly dependent on science and technology should still have taken it neither to its heart nor its mind. At present only one member of the Cabinet has a science degree; and only a handful of the 651 members of the House of Commons have scientific or engineering experience or qualifications. C.P. Snow's famous description of the two un-interacting cultures may be out of date in some respects. Our real rulers today are business managers, accountants and market operators. Such people may be indispensable but they are not known for the longer view. So when decisions need to be made – as they do every day – on matters of scientific or technical policy, alarming mistakes can be, and often are, made. Those who do not understand the wider scientific background can all too easily be bamboozled by self-interested experts, and rarely see the long term implications of their actions.

Likewise, the ability to judge environmental issues in the face of uncertainty should form part of every citizen's intellectual equipment. Our education system needs to recognise the interconnectedness of environmental issues and the complexities of their relationships. Students must be able to integrate the perspectives gained from different disciplines into a coherent view of the world. That will mean teaching the teachers and examining the examiners. It also means holding up a mirror to ourselves.

Conclusions

In marrying the disciplines, we have to make sure that consummation in the sense of action, follows. To bring it about soon will need a rare combination of factors: leadership from above; pressure from below; and some catastrophe to jerk people out of their inertia, and give new focus to politics. It could be extreme weather events (droughts or floods), a major food crisis, linked perhaps to genetically modified organisms going wild, millions of refugees on the march, threats to health through new or adapted micro-organisms (AIDS or Creutzfeldt Jakob Disease), or spreading social and economic breakdown.

Scientists have major responsibilities. In spite of recent disasters, they – with doctors – are the high priests of our society. They have to think big and wide, and accept the risks of trying to look ahead. In short they must be brave. The future is unlikely to be upwards and onwards. It could as easily be sideways or downwards. It will certainly include the unexpected. But the framework of science is large and flexible enough to embrace almost anything. As E.M. Forster said: "Only connect".

ICE AGE

Whiteness falls, cold augury of death.
Bright axe gleams and autumn's red head
rolls. Trees stand stiff and motionless,
blindfolded by snow, manacled in ince.

Solitary leaves hang reprieved,
camouflaged by light and thin as medals.
Winter hones its blade – poised, punctual,
exact – to execute the year.

Old tinker, living rough as pony might
at edge of field (his chosen home a tipsy
shack of branches, straw and sacking)
lay gripped in winter's vice; and died of cold.

I drive past, safety-belted, distrustful
of treacherous road, and miss his rude
dismissive fingerings of smoke –
the reassurance of his strenuous indifference.

Compromised, ineluctably secure, I peer
and steer along the squeaking ruts of snow
to year's and journey's end,
to frozen winter and my measurement of time.

Ken Morrice

THE EBB AND FLOW OF GEOLOGY IN BRITAIN

NORMAN E. BUTCHER

Mr Norman E. Butcher was Staff Tutor in Earth Sciences for The Open University in Scotland 1971–92; previously Lecturer in Geology in the University of Reading. He has served as Honorary Secretary and President of the Edinburgh Geological Society. He is currently Chairman of Lothian and Borders Regionally Important Geological Sites Group.

1997 brings the Bicentenary of the death of James Hutton (1726–1797) and the birth of Charles Lyell (1797–1875), two of the foremost figures in the history of geology throughout the world. It thus provides an opportunity to review, however briefly, the development of this particular science over the last 200 years. This essay attempts, however tentatively, to chart the ebb and flow of geology in Britain since the time of Hutton's death in Edinburgh in 1797.

Geology is not an easy science to get to grips with, yet it is everywhere around us wherever we might be. We cannot escape it, nor should we ignore it, since it determines so much of our civilisation. Sir Charles Fraser surely spoke for the majority of people when he reportedly said: 'I've always found geology difficult to understand.' (*Scotsman* 19 July 1996). This is a great pity, but there are good reasons for this common view. They are largely rooted in the history and development of the science over the last 200 years.

Geology before Hutton

There is a vast but fragmentary body of knowledge concerning the Earth gathered in the thousands of years preceding the time of Hutton. Although it did not formally exist as a science, Henry Faul and Carol Faul have argued that 'Geology began when early man first picked up a stone, considered its qualities, and decided that it was better than the stone he already had.' Particularly throughout Europe and the Mediterranean region, but also elsewhere in the world such as ancient China, there are well-documented examples of man's early use of geological materials and his attempts to understand his natural surroundings.

In Britain, a comprehensive collection of geological materials for academic study was established by John Woodward (1665–1728) in Cambridge, where it still exists in the Sedgwick Museum. Woodward's *Essay Toward a Natural History of the Earth*, published in 1695, concentrated on the Universal Deluge. One of the most fanciful and popular books produced in the late seventeenth century was that by the Cambridge divine, Thomas Burnet (1636–1715). First published in Latin in 1681, the later English versions were entitled *The Sacred Theory of the Earth*, an attempt to reconcile an imperfect Earth with Genesis. David Daiches, author and literary critic, has published the following reflections on Burnet's doctrines in the *New Yorker* (1956):

Sing a Song of Symmetry

    Burnet! You wring my heart full sore,
    I had not realised before
    How crooked, twisted, bent, and curled
    Is this our poor, distorted world.
    I look below and at my feet
    I see misshapen grasses meet.
    Above me, in the endless sky,
    The malformed clouds float sadly by.
    I eye my limbs. What grim surprises!
    Fingers and toes of different sizes!
    I don't like this at all; it rankles.
    My calves are thicker than my ankles.
    How can such disproportion please?
    And what is one to make of knees?
    No comfort, none, I must confess
    The human form is just a mess;
    And animals – let me be terse –
    Are, on reflection, even worse,
    While sky and meadow, field and tree
    Are just as ugly as the sea.

    Enough! I shut my burdened eyes
    And do my best to visualise
    A world where all is neatly framed
    and Burnet need not be ashamed.
    Ah, in this ideal world I see
    Nothing so monstrous as a tree,
    But o'er her surface Nature drapes
    Some pleasing vegetable shapes;
    No rudely disproportioned storms
    Distort their geometric forms;
    If any breeze their limbs inclines,
    They keep to parallel straight lines.
    The world of living creatures shares
    The zeal for perfect rounds and squares.
    The worm that wriggles in the soil
    Moves neatly in a single coil.
    Well-patterned dogs with cube-shaped muzzles,
    Checked with squares like crossword puzzles,
    Symmetric pigs whose shapely hams
    Form perfect parallelograms –

Such ordered creatures play their part
In boosting their Creator's art.
How trim the patterned ocean strand
In perfect squares of yellow sand!
I note the rocks that bound the seas –
Triangular (isosceles).
Neat fishes as they swim adduce
The square on the hypotenuse,
And I observe the shapes of clams
Like Euclid's finest diagrams.

A nymph approaches – creature fair!
How straight and even is her hair!
Her eyes two perfect circles are;
Her bosom is rectangular.
Am I delighted? I am not!
– Burnet, you talk the damnedest rot.

James Hutton, Founder of Modern Geology

There can be little doubt that, of the four great figures of the Scottish Enlightenment (Adam Smith (1723–1790), David Hume (1711–1776), Joseph Black (1728–1799), James Hutton (1726–1797)), the name of Hutton is still the least known. The label attaching to his name was only fixed firmly to the wall in the Greyfriars Kirkyard in Edinburgh where he is buried at the time of the 150th anniversary of his death in 1947.

Born in Edinburgh in 1726, the son of William Hutton, merchant and City Treasurer, and Sarah Balfour, James Hutton attended the High School and Edinburgh University before being apprenticed to a lawyer. His early interests in chemistry and anatomy brought an abrupt end to his apprenticeship and, after re-entering Edinburgh University as a medical student, he moved to Paris for two years and then to Leyden, being awarded his MD in 1749. Returning to London and Edinburgh, he never practised as a physician but instead took up farming, first in Norfolk and later in Berwickshire where he had inherited two small farms from his father. In Edinburgh, he had gone into partnership with John Davie, producing sal ammoniac from coal soot. He travelled extensively through southern England and northern Europe in the 1750s. In 1764, Hutton toured N.E. Scotland with his friend, George Clerk Maxwell. Hutton moved back to Edinburgh permanently at the end of 1767, joining the Philosophical Society and in 1783 co-founding the Royal Society of Edinburgh. He thus became an integral part of the Scottish Enlightenment at that time.

Hutton's first publication was a small pamphlet in 1777 dealing with the distinction between Scottish coal and English culm for taxation purposes. He was also involved in the Forth and Clyde canal project. His first paper on the *System of the Earth, its Duration, and Stability* was read to the Royal Society of Edinburgh in 1785. The full paper appeared in the first volume of the Society's *Transactions* in 1788 whilst, in book form, his *Theory of the Earth, with Proofs and Illustrations* appeared in two volumes in 1795. This was in fact his fourth book, the first being published in 1792, the second and third both in 1794. These other books cover non-geological subjects and demonstrate the extraordinary range of Hutton's genius. Two volumes of manuscript, his *Elements of Agriculture* remained unpublished at the time of his death in 1797, whilst a third volume of Hutton's *Theory of the Earth* was only published in 1899.

Although it is still not entirely clear exactly how James Hutton got interested in the Earth, it seems that his natural curiosity combined with his extensive travels enabled him to gather observations wherever he went. His approach was to collect specimens of rocks and he clearly was most interested in their chemical aspects. Between about 1768 and 1785 he was largely concerned with the nature and origin of basalt, but with his discovery of granite veins in Glen Tilt in September 1785, the nature and origin of granite became a major topic for him. In 1787 and 1788, he discovered the three unconformities in Arran, at Jedburgh and at Siccar Point on the Berwickshire coast for which he is renowned. Hutton himself did not make drawings of what he observed on his field trips, but he was often accompanied by his friends, notably the artist John Clerk of Eldin, also keenly interested in geology. In the event, John Clerk of Eldin's sketch of the Jedburgh unconformity was the only engraved drawing of a locality published in Hutton's *Theory of the Earth*. His many other drawings were only published in 1978 following their discovery at Penicuik House near Edinburgh in 1968.

Without the visual impact which Clerk of Eldin's illustrations would have brought to Hutton's *Theory*, it is perhaps not too surprising that Hutton and his writings fell into some obscurity after his death. However, put simply, Hutton's enormous contribution was to realise the immensity of geological time and the cyclicity of geological processes. He acknowledged for the first time that present day processes operated in the past and were sufficient to account for what could be observed. He correctly identified heat as providing the energy for driving geological processes.

Geology after Hutton

It is important to remember that, unlike his friend the chemist Joseph Black, Hutton held no Chair in the University of Edinburgh, and therefore there was no body of students to carry his ideas forward. It is true that another of Hutton's associates, the Professor of Mathematics, John Playfair, did his best to promote the Huttonian philosophy after Hutton's death, but the appointment of the Leith-born Robert Jameson (1774–1854) to the Chair of Natural History in the University of Edinburgh in 1804 heralded 50 years of the opposing Wernerian view of the origin of rocks. Jameson had studied briefly with Abraham Werner in Freiberg and in Edinburgh he established the Wernerian Natural History Society. With his Museum set up in what today is the Talbot Rice Art Gallery in Old College and with a considerable number of people coming to Edinburgh to study with him, together with the members of his Society, there is no doubt that Jameson's influence was considerable. Living in Moray Place in the New Town of Edinburgh, in contrast to Hutton living with his three sisters on St. John's Hill in the Old Town, Jameson seems to have been a leading member of Edinburgh society in the first half of the 19th century. However, some such as Charles Darwin, briefly a medical student in Edinburgh in the 1820s, found Jameson's lectures extremely dull and boring. His Wernerian "Geognosy" and his Society petered out near the end of the 1830s.

Meantime, in England

Working nearly 50 years later than Hutton in Edinburgh, William Smith (1769–1839), Father of English Geology, adopted a significantly different approach from that of Hutton in the country around Bath and Bristol. Born in the village of Churchill in Oxfordshire, he was trained by a land-surveyor called E. Webb in Stow-on-the-Wold. Smith was involved in surveying the line of the Somerset Coal Canal, and in the course of his work discovered the principle of stratigraphy, recognising that each stratum contained its characteristic assemblage of fossils. He also recognised the general easterly inclination of the strata in the west country and began to make a map of the distribution of the strata, using a unique method of colouring the outcrops on the map. Spreading out from the area he initially surveyed, by 1815 he had amassed sufficient information to produce his great map of England and Wales, with part of Scotland. He had moved from Bath to London, dedicated his great map to Sir Joseph Banks, President of The Royal Society, but the financial burden he incurred crippled him, spending some time in a debtor's prison. Like Hutton, he collected specimens but arranged them in his house off

Figure 1 *James Hutton 1726–1797 by Sir Henry Raeburn 1756–1823. Although thought to be one of the earliest known portraits by Raeburn, painted perhaps as early as 1778, it seems more likely to have been painted around 1795, judging from the pile of manuscript (surely of the* Theory of the Earth?*) and specimens on the table. Reproduced with permission of The Scottish National Portrait Gallery.*

the Strand in London in the order in which they occurred in the ground, thus displaying a remarkable three-dimensional grasp of the strata. Unlike Hutton, he failed to communicate to any great extent his findings in writing, preferring to communicate his results visually, especially at sheep-shearing and other agricultural events. "Strata Smith" became well known and from 1815 was assisted by his orphaned nephew, John Phillips (1800–1874), later Professor of Geology at Oxford. William Smith went on to produce a series of geological maps for some 21 individual counties, forming a six-part Geological Atlas, between 1819 and 1824. For all his maps, uniquely hand-coloured, he used topographic base maps specially supplied by J. Cary of the Strand in London. All his maps are now quite rare. The only society Smith ever joined was the Bath and West Agricultural Society, but he attended the Annual Meetings of the British Association for the Advancement of Science, following its inception, largely due to John Phillips, in York in 1831. Smith was on his way to the Annual Meeting in Birmingham in 1839 when he died and is buried in St. Peter's graveyard in Northampton.

Charles Lyell, populariser of geology

Often referred to as an 'English geologist', Charles Lyell was born in Scotland in the year that Hutton died, 1797, at the family estate of Kinnordy, just outside Kirriemuir in Forfarshire. Apparently, his parents did not like Scotland and, in 1798, took a 14 year lease on Bartley Lodge, a substantial house with 80 acres just west of Southampton. The lease was renewed for another 14 years in 1811, so Charles Lyell's boyhood years were largely spent in Hampshire. Attending schools in Ringwood, Salisbury and Midhurst, he was dogged by illness but was considerably influenced by his father's library at home at Bartley Lodge, reading the first English textbook on geology by Robert Bakewell, published in 1813. Lyell entered Exeter College at Oxford in February 1816 to read law but attended William Buckland's mineralogy course in 1817. William Buckland (1784–1856), successively Reader in Mineralogy and Reader in Geology at Oxford, enjoyed a considerable reputation for his geological teaching at this time. Also in 1817, Lyell started the extensive tours from which he was to gain so much experience, visiting Scotland with his father. They called on Robert Jameson in Edinburgh and examined Calton Hill and from Kinnordy went to see Staffa, then an object of much curiosity with its basaltic columns. In the following year, Lyell visited Paris, Geneva and Chamonix. In 1819, Lyell became a Fellow of the Geological Society in London and read his first paper to that Society in 1824. Although he practised briefly in London at law and also held briefly a post at King's College, Lyell was to devote himself completely to geology and to his writings, developing an elegant style. In 1827, he began writing a book 'in confirmation of ancient causes having been the same as modern ones.' An ambitious tour in 1828 with Roderick Murchison and his wife took them to Padua, thereafter Lyell alone working his way south as far as Sicily, returning to London early in 1829.

The first volume of Lyell's *Principles of Geology, Being an Attempt to Explain the Former Changes of the Earth's Surface, By Reference to Causes Now in Operation* was published in London by John Murray on 24 July 1830. Charles Darwin, the Edinburgh drop-out, but later at Cambridge, took a copy with him on his five-year voyage round the world in The Beagle. The second and third volumes of Lyell's *Principles* were published in 1832 and 1833 respectively and by the time of his death in 1875, this milestone title in the history of geology went through 12 editions. Lyell wrote several other books, including his *Travels in North America* (1845) based on his first visit there in 1841 when he was invited to give the Lowell Lectures in Boston. Charles Lyell was regarded by the American geological

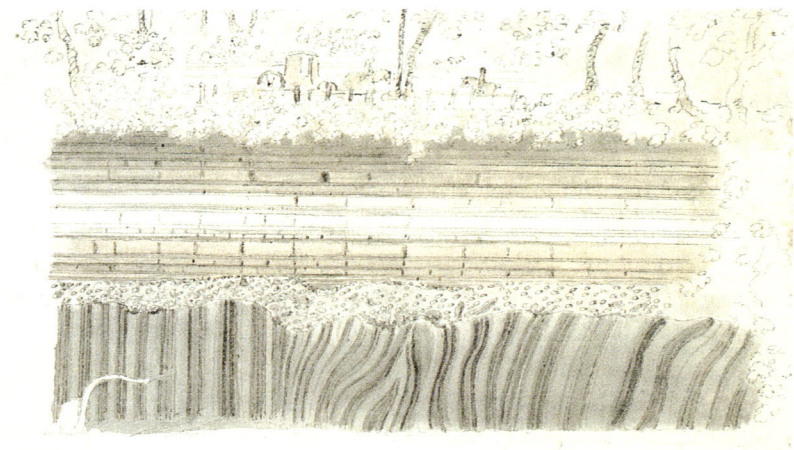

Figure 2 *The unconformity in the bank of the River Jed half a mile south of Jedburgh in the Scottish Borders. Drawn by the artist John Clerk of Eldin in 1787, it shows the horizontal Upper Old Red Sandstone resting directly on the eroded surface of vertical and folded Silurian greywackes and shales. Reproduced with permission of Scottish Academic Press and Sir John Clerk, Bt. of Penicuik.*

community as the leading British practitioner of the science. The Present as the Key to the Past received royal approval by the knighthood conferred on Lyell in 1848. Geology became the foremost science in Victorian Britain. On his death, Lyell himself was buried in Westminster Abbey in London. Not for him the obscurity conferred on Hutton in Edinburgh.

The rise of geological institutions in Britain

The formation of the Geological Society in London in 1807 was a vital force in the rapid development of geology in the first half of the nineteenth century in Britain. It is the oldest such society in the world and with its meetings and publications, acted as a focus of debate on issues of the day. Its membership included not only the leading exponents of geology but also amateurs such as Lyell's father-in-law, Leonard Horner, an Edinburgh merchant who played a key role in setting up many institutions. It was not long before other geological societies were set up in Britain, notably the Royal Geological Society of Cornwall in Penzance (indeed, the only geological society carrying the appellation 'Royal') in 1814 and the Edinburgh Geological

Figure 3 *Sheet XI of the fifteen separate sheets forming William Smith's* Delineation of the Strata of England and Wales, with part of Scotland, dedicated to Sir Joseph Banks, Bt. *and published by J. Cary of the Strand, London 1815. At a scale of five miles to one inch, this superb geological map shows Smith's unique method of colouring the strata. From the copy of the map which formerly belonged to the Bath and West Agricultural Society.*

Society in 1834. Locally-based amateurs dominated these other societies.

With the formation in York in 1831 of the British Association for the Advancement of Science, already referred to, the leading geologists of the day had an additional forum at the Annual September Meeting, held in a different town or city each year. One such notable Meeting was that held in Glasgow in 1840, after which Charles Maclaren, Editor of *The Scotsman* (himself an amateur geologist) announced in that paper the discovery of the former glaciation of Scotland by Louis Agassiz of Neufchâtel.

Arguably the greatest force for the advancement of geological knowledge of Britain was the establishment of the official Geological Survey in the 1830s. Initially, in England and Wales only, but with interesting and important links with Ireland also, the Survey was

Figure 4 *Charles Lyell 1797–1875. Calotype by D.O. Hill 1846. Reproduced with permission of Lady Lyell.*

extended into Scotland in the 1850s. The creation of the Geological Survey, the world's oldest continuously operating publicly-funded such organisation, was due to the extraordinary energies of one man, Henry Thomas De la Beche.

Henry De la Beche (1796–1855) was born in London, attended school mostly in the west country and entered the Military School of Great Marlow from which he was dismissed in 1811. He then put his flair for drawing to good use in the new science of geology. He had inherited his father's Jamaican estate of Halse Hall, in Clarendon, in 1801 and in 1827 published in London the first account of the geology of part of Jamaica based on his second visit there in 1823–24. With experience of geological mapping under his belt, in March 1832 De la Beche wrote to the Board of Ordnance offering to survey geologically the ground covered by eight Ordnance Survey one-inch sheets in south-west England. The work done, De la Beche was appointed to a

Figure 5 *The ruined columns of the Temple of Serapis at Puzzuoli near Naples. Visited by Lyell in 1828, he argued that the borings of marine organisms in the columns showed that the land had both been depressed and later elevated since classical times. This view formed the Frontispiece to Lyell's* Principles of Geology *published in 1830. In later editions, it was also embossed in gold on the front cover of this famous book published by John Murray of Albemarle Street, London.*

permanent post at the head of a new Ordnance Geological Survey of England and Wales. The first publication of the new Geological Survey was De la Beche's massive 648 page *Report of the Geology of Cornwall, Devon and West Somerset*, produced in February 1839. Not only did De la Beche become the leading practitioner of the observational science of geology in geological mapping, but also in its application 'to the useful purposes of life, or, in other words the mineral wealth of the country.' He pressed the case to Government, successfully, for the establishment of a Museum of Economic (Practical) Geology, the Mining Records Office and the Royal School of Mines. All three were founded in London between 1837 and 1851.

Nineteenth century development
It was perhaps inevitable that as the new science of geology rapidly developed in the nineteenth century, so it became more subdivided, each branch of the subject becoming more and more burdened with

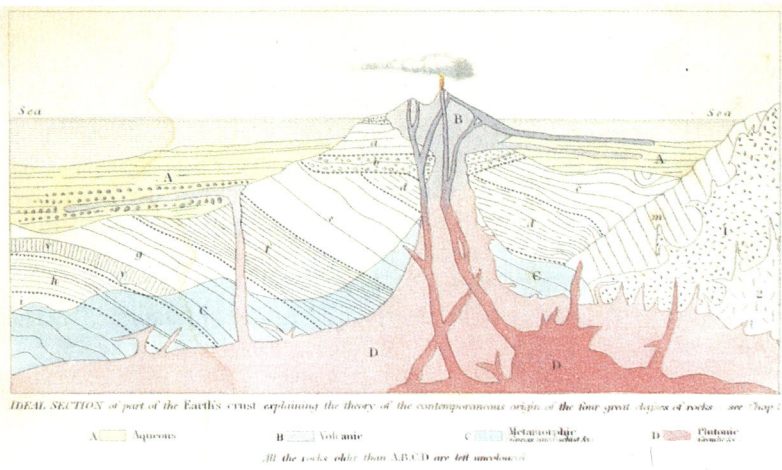

Figure 6 *The frontispiece to Charles Lyell's* Elements of Geology *published by John Murray of London in 1838, showing an ideal section of part of the Earth's crust.*

its own terminology. Geology became essentially a descriptive science. So, for example, the study of rocks and their constituent minerals became petrology with a bewildering array of rock and mineral names for students to work through and learn. Collections of specimens were established, in national and local museums and in those universities which established, as most did, Geology Departments. In particular, the application of the optical polarising microscope with the important use of the prism of Iceland Spar, the mineral calcite, invented by William Nicol (1768–1851) of Edinburgh, opened up a completely new field of study in optical mineralogy.

Similarly, the nineteenth century saw the growth of palaeontology, the study of the morphology and classification of the vast array of fossils being found in sedimentary rocks, both invertebrate and vertebrate animal fossils as well as plant fossils. Publication began of huge illustrated monographs containing systematic descriptions of the various groups of fossils.

Based on field observations, geology was essentially a historical science, as Lyell had emphasised. One of the main pre-occupations of the nineteenth century was the establishment of stratigraphy, the succession of rocks, especially sedimentary rocks, through geological time. This led, as in the case of the setting up of the Devonian system in south-west England, to some spectacular controversies, in which the participating geologists were often caricatured as warring combatants.

With the extraordinary single-handed achievement of William Smith as example, the nineteenth century saw the visual language of geology expressed above all in the production of geological maps. As F. J. North, Keeper of Geology in the National Museum of Wales, expressed it in 1928: 'The geological map may, therefore, be regarded as the dynamic force in Geology ...' The art of the geological surveyor lay, above all, in an appreciation of the three-dimensional nature of rocks as they occur in the earth, together with an appreciation of geological time. These remain the main obstacles to understanding of geology to be overcome by the uninitiated.

Twentieth century stagnation

It is significant that it was not until the 1960s that the first stirrings of active revolution occurred in the geological community in Britain. Up to that time, developed geology had reached a gentle state of stagnation. The image of the geologist was famously portrayed in the Schweppes advertisement. The general public was blissfully unaware of the science of geology, often simply confusing it with archaeology.

Figure 7 *The geological observer observed. A sketch by Sir Henry Thomas De la Beche, Director-General of the Geological Surveys of the United Kingdom, as one of the several woodcut illustrations accompanying his first article in Volume I of the Survey's Memoirs published in 1846. The section at Portishead near Bristol shows horizontal Dolomitic Conglomerate resting unconformably on vertical Old Red Sandstone.*

The ebb and flow of geology in Britain

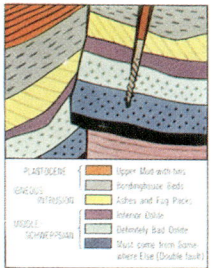

Figure 8 *Nature Watching in Schweppshire. No. 4 Geology Watching. A caricature of the geologist at work, appearing in the Schweppes Advert series in 1960.*

Figure 9 *Arthur Holmes 1890–1965. Regius Professor of Geology, University of Edinburgh from 1943 to 1956, Holmes was previously Professor at Durham University where he began his seminal studies on the age of the Earth.*

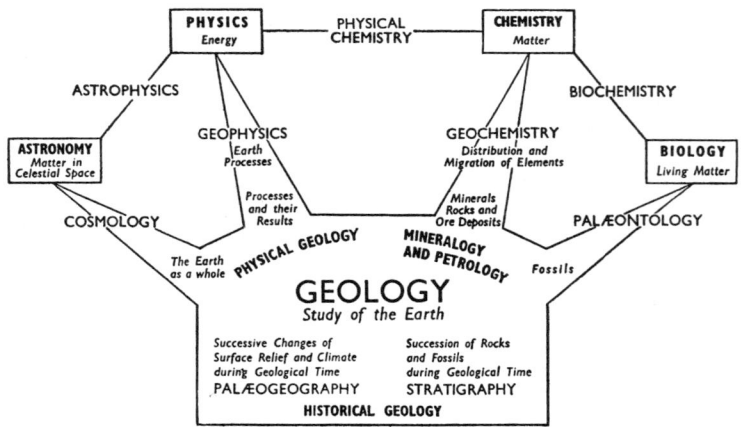

Figure 10 *Subdivisions of the science of geology and their relations to other sciences. This diagram formed Figure 1 in the classic textbook by Arthur Holmes, Principles of Physical Geology, first published in September 1944.*

Figure 11 *The cover of* Understanding the Earth, a Reader in the Earth Sciences, *comprising 25 chapters written by specialists on a wide range of topics. Published in 1971 as the Set Book for The Open University's Science Foundation Course, it was written, edited and published by Artemis Press in the remarkably short time of nine months.*

Figure 12 *Ian Gass 1926–1992. Professor of Earth Sciences in The Open University from 1969 until 1991, Gass was formerly at the University of Leeds. He was the first geologist to recognise that the island of Cyprus contains a fragment of old ocean floor, caught and pushed up between the colliding continents of Africa and Eurasia.*

The reasons for this stagnation are not entirely clear, but a major factor must lie in the way the science had developed into a number of quite distinct subdivisions, each with its own devotees. In universities particularly, where geology was largely preached, this meant that each specialism was simply carried forward by the next generation of students. At Cambridge, this extreme specialisation meant quite separate Departments with little or no communication between them!

The revolution

In the early 1960s, a group of dissidents, dissatisfied with the way the Geological Society in London was being run, met regularly at Schmidt's restaurant in Charlotte Street after formal meetings at Burlington House. In 1964, the first attempts were made to start a professional body for the science, the lack of public awareness of geology being expressed in some notable case histories, for example, in the construction of the M6 motorway near Keele in Staffordshire.

In 1965, the Geological Survey transformed itself into the Institute of Geological Sciences, a constituent body of the Natural Environment Research Council. In 1969, the international weekly scientific journal, *Nature*, published a short anonymous article which launched a scathing attack on the state of geology in Britain at that time.

Although the revolution in the science of geology, leading to Earth Sciences, was a world-wide phenomenon culminating at the end of the 1960s with the birth of plate tectonics as an all-embracing, unifying theory, one of the main architects in Britain of the revolution was the reclusive Arthur Holmes (1890–1965). Holmes, Professor of Geology at Durham University until 1943, first published a little book in the Nelson Classics series at 1s 6d in February 1927 on *The Age of the Earth*. He had become interested in the topic even as a schoolboy in Newcastle-upon-Tyne when his physics teacher encouraged him to read Lord Kelvin's papers. Kelvin's estimates took no account of natural radioactivity and Holmes took up with enthusiasm at Durham the task of constructing a geological time scale. Whilst fire-watching at Durham in the second world war, Arthur Holmes began to write a text book, *Principles of Physical Geology*, also published by Thomas Nelson and Sons Ltd, in September 1944, destined to become a classic. Holmes had transferred to Edinburgh University in 1943 as Regius Professor of Geology. A prolific author, Holmes had speculated as early as 1928 on the role of convection within the earth. The penultimate sentence of his great book published during the last war read: 'Meanwhile, it would be futile to indulge in the early expectation of an all-embracing theory which would satisfactorily correlate all the varied phenomena for which the earth's internal behaviour is responsible.' Every schoolboy now should know that continental drift and sea floor spreading led directly to plate tectonics at the end of the 1960s.

Dynamic earth sciences

The first University Department in Britain to change its name from that of 'Geology' to 'Earth Sciences' was Leeds, in the late 1960s, the feeling being that 'Geology' no longer adequately covered the vast range of the subject matter involved, especially with the growth in this century of geochemistry and geophysics. By an extraordinary coincidence, 1969 saw the setting up in Britain of The Open University with Ian Gass (1926–1992) from the Leeds Department appointed as Foundation Professor of Earth Sciences in April of that year. With phenomenal energy and vision, Gass rapidly built up from nothing a Department of world renown, offering courses in modern

Earth Sciences to large numbers of adult students studying part-time throughout Britain. The Department also undertook impressive research projects scattered round the world. Perhaps the greatest contribution to the geological community world-wide was the publication, in 1971, of *Understanding the Earth*, thus bringing the new knowledge of the earth to a vast audience. With Ian Gass's death from a second stroke in October 1992, and with other problems, the educational spivs gained a stranglehold on The Open University.

Also, beginning in the 1980s, University Departments were re-organised after a Review by Professor Ronald Oxburgh who had successfully unified the separate Departments at Cambridge. Some smaller Departments of Geology were closed altogether, creating fewer and larger Departments. But the later creation by Government of new Universities from the former Polytechnics added several Departments which had escaped re-organisation.

The Institute of Geological Sciences changed its name yet again, this time to the British Geological Survey, and at the time of writing the future of this largely publicly-funded organisation remains unclear. BGS has, however, begun a drive to communicate the nature of its work more and more to the public generally (but see Appendix).

With undoubted growing interest in the environment generally, and topics such as climatic change specifically, it is being realised that not only is the Present the Key to the Past, but the Past is the Key to the Present and possibly our guide to the Future. Only the study of the Earth and the rocks it contains will reveal the record of long-term environmental change.

In 1997, we do well to celebrate the Bicentenary of the death of Hutton and the birth of Lyell and acknowledge the foundations they secured for this most important science.

Appendix
The Government has now announced that the BGS should remain in the public sector (30 January 1997).

SCIENCE FESTIVAL (3.18×10^8 ms^{-1})

dear festival
you be my light and illusion
my three point one eight

i go walking down road this morning
to work for you
when my steps feel like sneezes
and i get how some people fast
and some go slow
where you just is
you my three point one eight

i been sitting in your lectures
not getting round sides of you
not getting to push away
sitting in your illusions
of day bigger than night
of the pretty girl and hag

and i been seeing kids
taking the doors at adam house
like they dive into big swimming pools
all splashed and shouting in science
while i yawn for help

dear festival
your after-image is pub
round by round
it slow dying and fading
till i can see smiles
till i be ready for morning
and my three point one eight

Michael Forester